TEACHING MATHEMATICS

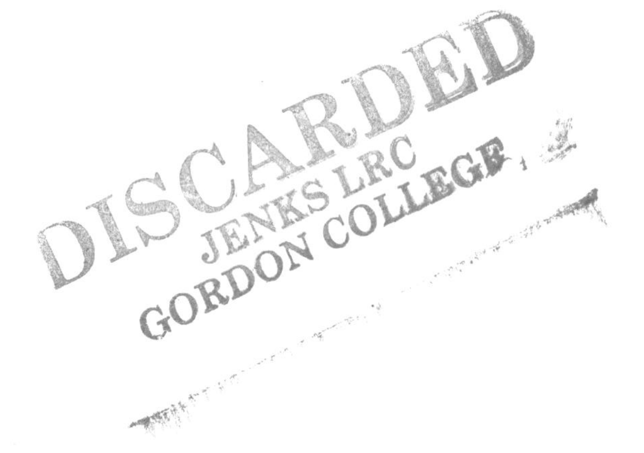

OTHER BOOKS OF RELATED INTEREST

THE TEACHING OF HISTORY
Dennis Gunning

TEACHING CHILDREN IN THE LABORATORY
Joan Solomon

TOPIC TEACHING IN THE PRIMARY SCHOOL:
Teaching About Society Through Topic Work
Stella Gunning, Dennis Gunning and Jack Wilson

MIXED ABILITY TEACHING
Edited by Margaret Sands and Trevor Kerry

TEACHING ENGLISH
Tricia Evans

STARTING TEACHING
R. Honeyford

Teaching Mathematics

Edited by
MICHAEL CORNELIUS

CROOM HELM
London & Canberra

NICHOLS PUBLISHING COMPANY
New York

Croom Helm Ltd, Provident House, Burrell Row,
Beckenham, Kent BR3 1AT
Croom Helm Australia Pty Ltd, 28 Kembla Street,
Fyshwick, ACT 2609, Australia
Reprinted 1984

British Library Cataloguing in Publication Data

Teaching mathematics.
1. Mathematics – Study and teaching
I. Cornelius, Michael
510'.7'1 QA11

ISBN 0-7099-0714-1

First published in the United States of America 1982 by
Nichols Publishing Company, Post Office Box 96,
New York, NY 10024

Library of Congress Cataloging in Publication Data
Main entry under title:

Teaching mathematics.

1. Mathematics – Study and teaching.
I. Cornelius, Michael.
QA11.T348 510'.7'1 82-6404
ISBN 0-89397-137-5

Printed and bound in Great Britain by
Biddles Ltd, Guildford and King's Lynn

CONTENTS

PREFACE

Today's teacher of mathematics is faced with many problems, much criticism and a wealth of suggestions about what should or should not be done in the classroom. At a time when there is a shortage of qualified teachers of mathematics, schools are under many pressures in areas where there is often no general agreement about an appropriate policy or method: for example, the use of computers or calculators in the classroom, the use of workcards or individualised learning methods, the teaching of pupils of extremely low or high ability, or the organisation of children into sets or groups of mixed ability. Teachers and would-be teachers are bombarded with ideas and materials in many guises but often have little time or access to informed advice to help them attempt to evaluate and assess available resources.

This collection of papers, written by ten contributors with a wide experience in mathematics teaching, aims to provide teachers at all levels with ideas, background and thoughts on a number of important aspects of mathematical education. Rarely can there have been a period of such intense debate on the teaching of mathematics as has been witnessed in recent years. The publication of the report of the Committee of Inquiry into the Teaching of Mathematics in Schools (the 'Cockcroft' report)[1] has provoked wide discussion and the contributions which follow examine some contemporary issues in school mathematics in the light of this and other recent reports. It is impossible in a single volume to cover *all* important aspects of mathematical education but this book seeks to stimulate thought on at least some of the vital current issues relating to the teaching of mathematics at all levels in schools.

There are no quick and easy solutions to most of the problems in mathematical education. The chapters which follow will, it is hoped, emphasise some of the diversity of opinion which exists and serve as a reminder that the individual teacher of mathematics faces an exciting, fascinating challenge and also bears an awesome and heavy responsibility.

Michael Cornelius

1. Throughout the book references to the Cockcroft Report are to the report of the Committee of Inquiry into the Teaching of Mathematics in Schools, entitled *Mathematics Counts*, (HMSO, London, 1982).

ACKNOWLEDGEMENTS

The O-level examination question on page 197 is reproduced with the permission of the Joint Matriculation Board. Figure 9.6 is reproduced with permission from Blackie and Son Ltd.

ABBREVIATIONS

(The following lists only those abbreviations not extended in the text).

APU	Assessment of Performance Unit
CSE	Certificate of Secondary Education
DES	Department of Education and Science
GCE	General Certificate of Education
	A-level: Advanced Level
	O-level: Ordinary Level
HMI	Her Majesty's Inspectorate
HMSO	Her Majesty's Stationery Office
ICME	International Congress on Mathematical Education
IMA	Institute of Mathematics and its Applications
LEA	Local Education Authority
MAB	Multibase Arithmetic Blocks
NFER	National Foundation for Educational Research
ORACLE	Observational Research and Classroom Learning Evaluation
OU	Open University
SMP	School Mathematics Project

1 CURRENT ISSUES AND PROBLEMS IN MATHEMATICS TEACHING

Rolph Schwarzenberger

1.1 Assumptions

Any discussion of mathematics and mathematics teaching must be based on conscious or unconscious assumptions about the nature of mathematics itself. For example that mathematics is:

- a set of techniques to be tested by examination,
- a body of knowledge to be learnt,
- a language using a particular notation,
- the study of underlying logical structure,
- an artificial game played by mathematicians,
- the construction of models useful in science,
- the calculating procedures needed for applications.

If any of these assumptions were by itself a sufficient description of mathematics then certainly there would have been no need for a government Committee of Inquiry into mathematics teaching. My own assumptions in this chapter are that all the statements quoted above have an element of truth, that the widely-differing assumptions and expectations concerning mathematics often lead to misunderstandings and disputes, that there are many built-in conflicts between different but equally valid aspects of mathematics which mean that problems of mathematics cannot admit any easy solutions, and that mathematics must be viewed as a complex social activity within the context of society as a whole.

Thus the current issues and problems of mathematics teaching reflect the issues and problems of society at large, and the burden of their solution cannot fall on mathematicians alone. It is for this reason that I begin with a brief, and perhaps idiosyncratic, summary of the social economic and cultural background to the Cockcroft Report.

1.2 Background to Cockcroft

We live in a consumer society which expects to buy ready-made goods off the shelf. Parents seek examination qualifications for their children, employers seek candidates with the requisite mathematical techniques, social scientists and physical scientists seek ready-made mathematical models, politicians and planners seek instant solutions, educational theorists seek clear-cut answers. It is hardly surprising that all should be disappointed that the goods purchased do not meet fully the expectations aroused. But disappointed expectations alone would not be sufficient to justify a Committee of Inquiry. More crucial perhaps is the persistence in our society of the belief that given any problem there must be a solution. This persistence is at first sight surprising in view of the daily evidence, from all modern societies, that problems are rarely solved and that solutions usually bring new problems. But the belief in instant solutions persists, fuelled particularly by the vested interests of those 'experts' who stand to gain from the solutions adopted.

Among these vested interests we must regretfully include the mathematical community. The role of mathematics in past scientific and technological successes, and the general acceptance that mathematical methods are central to the solution of all kinds of problems have rightly enhanced the importance of mathematics teaching. But mathematicians themselves have sometimes claimed too much and, in particular, have too glibly assumed that the problems of the mathematics curriculum itself can be solved as easily as problems in mathematics.

All these strands have contributed to the widespread belief that, if mathematics has not so far solved the problems of society, or if a large part of the population understands no mathematics beyond basic arithmetic, then the faults must lie with the mathematics syllabus. In the 1960s such beliefs led naturally to the mistaken notion that changes in the mathematics syllabus would be sufficient to remove the faults. In themselves the changes were for the better. At primary level the importance of play and discovery, and at secondary level the insistence on explanation and understanding of concepts, were important advances away from the rote learning of techniques. What was wrong was the haste with which changes were implemented – often by unwilling teachers ordered to change over to 'modern' by their head – and also the over-confident claims made on behalf of those changes.

This confidence was also reflected in the claims made during the

1960s and 1970s on behalf of the 'new' subject of mathematical education. The ICME conferences held in Lyon, Exeter, Karlsruhe and Berkeley, the launching of college and university courses on mathematical education, the increased research effort on problems of the teaching and learning of mathematics were all positive ventures; but the excessive claims as to the scientific standing of the new subject, the blind faith in the truth of some particular theories, and the introduction of quantitative and statistical techniques which imply a spurious accuracy of measurement of overall achievement, have all proved harmful to the development of good teaching and have contributed to a sense of disappointment and disillusionment within the profession. It is instructive to compare, for example, the introductory chapter by G.T. Wain, in a similar volume (Wain, 1978) published a few years ago, with the present chapter. It is not surprising that excessive claims made by some experts in mathematical education should have led in turn to excessive disillusionment and to the establishment of the Cockcroft Committee.

The theme of this chapter is that excessive disillusionment is as wrong as excessive claims on behalf of mathematical education. The basic problems and issues facing mathematics are as daunting and as challenging as those facing society as a whole, and in fact they are at root the same: how can we live in an age in which there are no longer any answers? Just as in the fields of politics and economics, and even in medicine, we are beginning to admit that many problems defy solution and that an easy answer which is attractive from one viewpoint will bring with it deleterious consequences from another viewpoint, so in mathematics teaching it is becoming clear that the lack of easy answers, the apparent lack of progress towards better understanding, and the way in which reform brings new problems in its wake, arise from features inherent in mathematics itself.

We may note that this is nothing new. It has been obvious to theologians and novelists for the last 30 years that we are entering an age in which there are no answers and in which beliefs in ready-made solutions – like beliefs in the millenium or in the emergence of the truly socialist state – should be viewed with suspicion as relics of an 'Age of Progress' long dead. So we need to focus on the special features of mathematics which cause this state of affairs: a state in which any 'solution' carries built-in tension between two extremes which are in conflict, and where a single easy answer is ruled out *a priori* by the nature of mathematics itself.

1.3 The Nature of Mathematics

At the beginning of this chapter I referred to some of the different meanings of 'mathematics' for different people. The very diversity of mathematics carries with it the consequence that different meanings may conflict: thought processes and understandings derived from one view of mathematics may disturb or even contradict those derived from a different view of the subject. We can classify the various views of mathematics under two very broad, and admittedly overlapping, headings.

The first includes all attempts at description, classification and understanding of relationships. Examples include understanding why algorithms work, devising geometrical or combinatorial models which help to explain social or physical phenomena, and the provision of algebraic or differential equation models for social or physical scientists. In the terminology popularised by Skemp (1971, 1976, 1979), the thoughts generated by this view of mathematics lead to *relational* understanding.

The second includes all attempts at calculation computation and problem solving. Examples include putting algorithms into effect, devising notations or programs which simplify a calculation, and making correct assumptions about a particular application. The thoughts generated by this view of mathematics lead to *instrumental* understanding.

This dichotomy between instrumental and relational is presented here in a slightly different form from that of Skemp. His terminology has proved a very useful way of describing the two different kinds of mathematical activity, and has given teachers a way of talking about an important distinction. It is an everyday experience that there are pupils who have ability at computation but cannot understand the principles, just as there are those who understand the principles but cannot perform quite simple computations. What is less useful, and even harmful, is the implication that relational understanding is preferable to instrumental understanding on the grounds that the former allows adaptation to new circumstances whereas the latter does not. A statement that relational understanding is 'better' than instrumental understanding can make sense only in respect to a particular piece of mathematics and a particular view of the nature of mathematics. In fact both relational understanding and instrumental understanding are desirable and, in certain contexts, vital. But they can very easily be conflicting: examples can be seen every day in the classroom but here are four in which a 'modern' syllabus has given priority to the understanding of general principles in a way which actually hinders the simultaneous development of efficient computation:

(1) The teacher takes great care to explain the concept of directed number and the notation $^{+}4$, $^{-}5$ etc. which leads to impossible sums like $+\ {}^{+}4 + {}^{-}5$, $-\ {}^{+}4 + {}^{-}5$ etc.
(2) The teacher takes great care – perhaps using special apparatus – to explain that multiplication of numbers has to do with the calculation of area; this is helpful for some purposes but not when learning multiplication tables or doing long multiplication.
(3) The teacher takes great care to explain $i = \sqrt{-1}$ as a transformation rotating the plane through $90°$; this helps to explain why $i^2 = -1$ but does not help in computations like $(3 + 4i)/(3 - 4i) = (3 + 4i)^2/25 = (-7 + 24i)/25$ when it is much more efficient to write $i^2 = -1$ without thinking too much about it!
(4) The teacher takes great care to explain the gradient $f'(x_0)$ at $x = x_0$ of a graph $y = f(x)$, perhaps as the slope of the tangent or perhaps as the best linear approximation to the function f; either way the pupil may actually be hindered by such thoughts when attempting to calculate the derivative of a complicated algebraic expression.

In all these examples the teacher has given priority to relational understanding, i.e. the thought processes necessary for the next stage of the mathematical development, over instrumental understanding, i.e. the thought processes necessary for efficient computation. There is nothing wrong in this so long as all parties realise what is happening and are aware of the conflict between the two ways of thinking.

The frequent conflict between relational and instrumental understanding which has been outlined above, is closely related to the conflicts which arise from the 'varifocal' nature of mathematics. Because the basic method of mathematics is to construct simplified models, there is always the choice as to the degree of simplification adopted. Paradoxically, the greater the degree of simplification the greater also is the degree of abstraction: the more a model is simplified by casting away variables which have significance in real life the more the user has the impression of an abstraction which is unrelated to reality. The success of a model depends entirely on hitting upon the correct focus: enough simplification to enable calculation and computation to take place but not so much that important variables are ignored. An analogy from recent literature may be helpful: the works of J.R.R. Tolkien include *The Hobbit* which recounts at full length a minor incident in the conflict described in *The Lord of the Rings*. Just as the whole of *The Hobbit* corresponds to a few pages of *The Lord of the Rings* so the whole of the latter work corresponds to a few pages of *The Silmarillion*. The focus in each case is different, but most critics would agree that the first book has the right combination of detail and abstraction for children, the second the right combination for adult readers and the third for very few readers at all. Choosing the correct focus is the hardest task of a mathematics teacher especially if faced by a class of pupils with differing needs and expectations: there may be no approach which provides the correct focus for every pupil in the class. The meanings of instrumental and relational make sense only with respect to a given focus; in fact the meaning can be completely altered as the focus changes since what is regarded as relational understanding at one level of simplification/abstraction may be regarded as instrumental understanding at the next level.

1.4 Choice of Subject Matter and Teaching Style

The discussion so far has been on the nature of mathematics in general and on conflict between relational and instrumental understanding. This conflict raises further psychological issues which are considered in the next section and in Chapter 4; these arise whatever the particular content of the mathematics curriculum and whatever the teaching style adopted. But the choice of content and the long and short term aims of the teacher may again produce tensions which cannot be resolved by any facile solution.

> What is the place of mathematics within the school curriculum as a whole? An optional subject for the gifted or a compulsory subject for all? A service to those subjects which make use of mathematics or a subject in its own right?
>
> For what do the pupils in a given class need to learn mathematics? To attain short term success in an examination or to reach long term understanding? For immediate applications or for later further education? To yield useful techniques or to pursue general logic development?

> What particular pieces of mathematics need to be taught? Those in an examination syllabus or those required for professional skills? Should the approach be algebraic or geometrical, numerical or combinatorial, with or without the use of microcomputers?
>
> Is the aim to teach mathematics as a set of answers to questions, or as a process of investigation? Is the algorithm to be specified by the teacher or discovered by the pupils? Is the approach determined by the availability of particular equipment or textbooks?

Simply to list such questions is enough to realise immediately that no straightforward answers will be possible. The teacher will have to make constant fine adjustments in response to the varying needs of different pupils, to the pressures from parents and headteachers, and to the demands of other teachers and administrators. Not only is the teacher unable to adopt clear-cut answers to such questions, but even worse clear-cut answers are impossible because the answers required by the various pressures are in conflict. The complex nature of mathematics as a subject means that the teacher of mathematics is not to be programmed by pre-packaged answers by experts but must steer sensitively and skilfully through a tangled maze of conflicting requirements.

Because these conflicting requirements arise in large part from the particular nature of mathematics, and even more from the social pressures within the school, the teacher of mathematics must look for help primarily to fellow teachers of mathematics. Most issues of *Mathematics in School* and *Mathematics Teaching* contain several articles aimed precisely at helping teachers in their daily resolution of some of the conflicts which I have described.

1.5 Psychological Issues

I have referred above to some of the conflicts which arise between different approaches to mathematics. My assumption is that these conflicts are not reconcilable and that therefore the task of the mathematics teacher is to select the particular balance of approach which seems most appropriate in a particular context. It follows that there are dangers in an approach which over-emphasises one aspect of mathematics to the detriment of other aspects and this is confirmed by work on the psychology of learning.

First there is the danger of emphasising the fact that mathematics consists of answers to problems. If the short-term goal is the 'right answer' then equally important long-term goals may never be achieved. There is a tremendous pressure on both teacher and pupil to take short cuts by rote learning which may rigidify the pupil's mind in set habits for a whole lifetime. This over-emphasis leads to 'instrumental understanding' in its extreme form which is hardly understanding at all because it cannot easily be adapted to new circumstances.

Secondly there is the danger of emphasising the fact that mathematics consists of a collection of abstract structures. There is a tremendous pressure on both teacher and pupil to ignore concrete examples and to regard mathematics as an artificial game played with meaningless jargon. This over-emphasis leads to 'relational understanding' in its extreme, and thoroughly perverted, form which is not understanding at all.

Both these dangers are well documented in the psychology of learning and occur in other subjects as well as in mathematics. But both have been propounded in all seriousness by advocates of particular approaches to mathematics. Even if these obvious dangers are avoided there remain further pitfalls for teacher and pupil. Just as any other subject, mathematics is something that people *do* as well as something that people *learn*. Most mathematicians, and most of those who use mathematics, regard the activity as more important than the knowledge. This means that the teacher must encourage play and discovery rather than passive recitation, must encourage comparison and discussion of different methods or answers, and must react to the different responses of each individual pupil. It follows that every pupil has a potentially different curriculum, and that the requirements of the pupil and of the class as a whole may well conflict. In fact the mathematics teacher is being asked to perform an impossible task: to achieve development of the individual mathematical activity of each pupil while at the same time achieving certain minimum goals with the whole class. No wonder that many teachers abandon the task and return to an approach which encourages passivity among pupils. Later chapters take up this point but it does perhaps need to be emphasised: that whatever the social pressures towards short-term goals, whatever the psychological pressures towards facts and rote learning, and whatever the difficulty of the task, no pupil will truly learn mathematics unless it becomes a skill which is used in personal mathematical activity. This activity need not be at a complicated or abstract level but it must exist; otherwise, so-called learning and understanding are meaningless. The striking contrast between the aridity of school mathematics for most pupils and the enthusiasm with which the same pupils accurately compute darts scores, football pool combinations and betting odds is surely not explainable simply by 'motivation' (whatever that means) but by the fact that in the latter examples mathematics has become a positive activity instead of a negative passivity.

This pitfall, of forgetting that mathematics is an activity, is linked to two common pitfalls regarding the nature of understanding. The first is to think that understanding consists of the correct recall of facts. Because mathematics is an activity as well as a collection of facts to be learnt it follows that understanding is context dependent. The teacher must recognise that pupils may hold contradictory views at the same time in different mental compartments. A few examples of this phenomenon may be helpful.

(1) In equations involving x, what is x? Sometimes it is a number, like 4, sometimes a quantity, like 4 cm, sometimes a 'variable' or 'unknown' (whatever these mean). Similarly, in sums about simple interest, we say that the interest rate I is 15 per cent in some contexts but I = 15 in other contexts (such as the formula $P \times I/100$). The same pupils will be asked to give different answers to the questions 'what is x?', 'what is I?'

(2) In the expression cos x, do you understand what x is? It is an angle measured in degrees (if the activity is trigonometry) or in radians (if the activity is calculus). The same pupils will give contradictory answers to the question 'what is x?' according to the activity in which they are engaged.

(3) Most pupils will simultaneously affirm that the decimal .333. . . (recurring) is equal to one-third, when the activity concerns fractions, but that the decimal .999. . . (recurring) is a little bit less than one, when the activity concerns whole numbers. They are surprised and confused when the contradiction is pointed out to them because they do not usually combine the two activities at the same time.

(4) Whenever an equivalence relation occurs as for example, in the statement that $^1/_2 = {}^2/_4$ (i.e. two different fractions represent the same number, or are equivalent) the mathematician will sometimes assert that the equivalent objects are the same, and at other times assert that they are different. There is no

contradiction because the assertion is not absolute but with reference to a particular context.

If these examples seem artificial you will be able to think of others from your own experience. The point is that a statement of the form 'my understanding of X is such and such' is meaningless unless the context or activity in which X occurs is made clear.

The second, closely related, pitfall concerning understanding is to forget that it is pupil dependent as well as context dependent. Every pupil in a class has a different past and different pattern of thinking. It is well established that when faced with a new fact – whether a mathematical technique or a mathematical concept – the first reaction is to accommodate it to existing patterns of thought. The result is the 'concept image' which will be different in each pupil and different, or even conflict with, the 'concept definition' which the teacher imagines has been taught. The work of Tall and Vinner (1981) on such questions has shown that the conflict can be considerable and can persist for long periods: hence the disillusionment which hits the university lecturer who marks the examination scripts of first or second-year students and finds that they have used definitions and techniques remembered from their A-level course instead of those taught in their university lectures. To decide how to present a new topic with maximum effect therefore requires careful diagnosis of the prior thought patterns of each individual pupil. Once again the teacher has an impossible task unless the class is unusually homogeneous.

These matters are taken up in more detail in Chapter 4. It will be seen that effective learning requires an upward spiral simultaneously breaking new ground and in harmony with previous experiences and with the aims of the pupil. It needs to be linked to positive activity and to be consolidated by the satisfaction and stimulus of success. Clearly this is an ideal very difficult to attain – there are so many things which can go wrong. Thus it follows conversely that conflict with previous experiences, lack of purpose, negative passivity and persistent failure will set up a downward spiral ending in the fear of mathematics so often displayed in the population at large. For the purposes of this chapter the most relevant consequence of the extreme difficulties in achieving effective learning is the overriding importance of the teacher's personality.

Because effective learning depends so much on states of mind in the pupil – previous concept image, confidence, aims, interests – more will depend on the personality of the teacher than on this or that syllabus or this or that style of teaching. Most people find mathematics books impossible to read without some form of teacher contact. Most people can instantly pick up nervousness or misunderstandings in the teacher's own mind. Most people will turn their mind off if they feel that the teacher has no interest or understanding of the way they think. So the net effect of the psychological issues discussed is to become aware that there are no simple recipes for effective learning and that the teacher of mathematics has a more crucial and more demanding task than most educational administrators will admit.

1.6 Social Issues

At the beginning of this chapter I mentioned the expectations aroused by the consumer society. The attitude of parents to teachers, like the attitude to clergymen and doctors, has become that of a consumer buying a product and expecting redress if the product is defective. Similarly employers expect the school system to produce a standardised school-leaver able to perform certain tasks and willing to conform to the requirements of a large organisation. Both parents and employers expect the product to be certified non-defective at the moment of purchase. The pressure on teachers and pupils to give absolute priority to the short-term goal of specified grades in A-level, O-level or CSE is immense. Because these grades are materially affected by decisions made and performance achieved in previous years, this pressure is felt not only in the 16+ or 18+ year but all the way down the school, so that even junior schools are affected. Whereas in the past the effect of the higher status of teachers was to allow the teacher to concentrate on long-term goals through play, experiment, discussion and discovery, today such activity is liable to be constantly interrupted by the pressure for short-term results which will satisfy the 'consumers': the parents, the employers or the children themselves.

Clearly this state of affairs, already deleterious to effective mathematical understanding, has been made even worse by the recent slide into recession and the failure of society to plan positively for structural employment. The price of failure to attain the short-term goals of CSE or O-level may now be the certainty of lifetime unemployment. Pressure to attain these short-term goals by whatever means is intensified, while the likelihood of future unemployment removes at a stroke the teacher's justification for the pursuit of long-term goals and encourages disinterest among the pupils. In fact the pressures on pupils to lose interest in mathematics are many sided. In a society in which personal freedom is being eroded, the superficial freedoms of the teenager are doubly attractive; the attempt to keep mathematics as a compulsory part of the curriculum, for pupils whose future mathematical experience will consist only of the simplest addition and subtraction, alienates many who would have the ability to understand mathematics if they could see any point in it. Things are made worse by the lack of properly-qualified mathematics teachers: the pressures on the one hand to provide compulsory mathematics for all, to provide A-level mathematics in every school, and to change over to 'modern' syllabuses whether or not the mathematics teachers welcomed the change, have combined with the pressures on the other hand to 'promote' mathematics teachers to the more vital tasks of counselling, timetabling, head of house, deputy head and head teacher, and the result is a very large number of teachers who are teaching a subject which they do not enjoy and which they do not fully understand.

Added to this gloomy picture are the problems caused by a doctrinaire approach to setting and streaming. The attempt to combine freedom of opportunity for pupils of all ability levels with a hierarchical subject like mathematics, in which each step cannot be understood unless first the previous steps are mastered, has led to the proliferation of work cards which encourage the bad teacher to avoid teaching altogether. These pressures are discussed more fully and more constructively in Chapter 5, but in the present context it is worth drawing attention to the severe problems caused by the power granted to the head teacher. On a complex matter which surely requires the professional judgement of an expert subject teacher it is possible for an authoritarian head teacher to dictate patterns of organisation totally unsuitable for the particular children involved. The anarchy caused by unlimited power given to individual heads is paralleled by the power given to the individual local authorities to decide such matters as age of transfer and pattern of sixth-form organisation. A child unfortunate enough to transfer schools or to move house is liable to have to change from one kind of mathematics to another for several years in succession.

At the very time when the mathematics teacher is assailed by all these social pressures and is most in need of enthusiasm, and even evangelistic fervour, an important source of possible succour has been removed. University mathematicians have widened the gap between current research in pure and applied mathematics and the secondary school teacher, when the latter in any case is too busy coping with all the other pressures to even attempt to keep up with the progress of the subject. It is not surprising that many teachers lose their enthusiasm for the subject and are tempted to teach by rote, or prefer promotion to administrative tasks which are better paid and easier to perform. Ten years ago there was perhaps more confidence that the newly-recognised subject of mathematical education would supply a neat overall theory within which clear guidance would be available to teachers at every level from nursery school to university. Today it is clear that the insights of mathematical education are diffuse and complex, that no easy generalisations fit all children or all social circumstances and that mathematics itself cannot be summarised in any neat package. The task of the mathematics teacher is harder than ever before: there is now more onus on the individual and less support from external sources.

At the risk of sounding unfashionably pious let me state explicitly the conclusion towards which this brief survey of social issues seems inexorably to lead. The greatest barrier to better mathematics teaching is not mathematical or even psychological. It is not a matter of social status or of better financial incentives. At heart it is spiritual: do we as mathematics teachers have the courage and enthusiasm to remain true to our calling at a time of such pressures within the job and such lack of faith in the future among society at large? Do we have the spiritual resources to cope without the bogus props of facile answers or naïve belief in progress? Can we keep our subject fresh and alive at a time when destructive forces in society are being unleashed? Have we the stamina to continue travelling in a country with very few signposts and with little or no assistance on the journey? Do we still believe in the importance and relevance of mathematics to the pupils whom we teach?

If we can answer 'yes' to these questions then all the mathematical, psychological, social and financial barriers to better mathematics teaching will be of little consequence. But if we cannot the future is bleak. Modern literature is full of warnings of this dilemma facing late-twentieth-century man. As a representative quotation which sums up the basic issue facing the mathematics teacher, take the letter written by the *Magister Ludi*, senior official of the Glass Bead Game which represents the supreme cultural achievement – combining and surpassing art, music and mathematics – of the imaginary country of Castalia, in the novel by Hermann Hesse written 40 years ago:

> Although I have endeavoured to serve with all my strength, the conduct of my office is (or seems to me to be) threatened by a danger which resides in my own person, although that is probably not its sole origin. At any rate, I see my suitability to serve as Magister Ludi as imperilled, and this by circumstances beyond my control. To put it briefly: I have begun to doubt my ability to officiate satisfactorily because I consider the Glass Bead Game itself in a state of crisis . . . Permit me to clarify the situation by a metaphor . . . Here I am sitting in the top storey of our Castalian edifice, occupied with the Glass Bead Game, working with delicate sensitive instruments, and instinct tells me, my nose tells me, that down below something is burning, our whole structure is imperilled, and that my business now is not to analyse music or define rules of the Game, but to rush to where the smoke is. (Hesse, 1972)

1.7 Conclusions

It may be that this survey of current issues and problems seems to the reader depressingly pessimistic. If the mathematics teacher labours under so many impossible pressures, and if the theory of mathematical education can offer very little by way of help and guidance, what hope is there for future improvement? Or even, if in the late-twentieth century we no longer believe in improvements, what hope for avoiding future deterioration?

The response must be on two distinct levels. The first is to achieve more general recognition – if necessary by the writing of depressive accounts of current issues and problems in books such as this – of the supreme difficulty of the mathematics teacher's job. The individual teacher alone can resolve the tensions and conflicts which arise from a particular mathematical topic with a particular pupil or class of pupils. No educational administrator or academic educational theorist can provide solutions instead; on the contrary the contribution of the administrator and theorist must be to build up the teacher's confidence, when things go right, and to accept a share in the blame when things go wrong (unlike the present situation where the teacher gets the blame when things go wrong, and the administrators, theorists and text-book writers demand the credit when things go right!).

It may be that the setting up of the Cockcroft Committee arose from a desire to find scapegoats for what has, in the eyes of politicians, gone wrong. Perhaps the hope was that teachers should be found in the role of scapegoat. Certainly there was a belief that the Committee of Inquiry might find the answer which had eluded so many for so long ('I expect you to report in two years but if you find the answer sooner let me know'). The outcome has been very different; the effect of the arduous sifting of evidence has not been merely to exonerate the teacher but to emphasise the extent of the teacher's responsibilities and the particular difficulty of the teacher's job.

The second level of response must involve the admission that, in an age without answers, we are all travelling alone: our need is not for signposts or even guide books but for self-confidence and self-reliance. The Cockcroft Report is a helpful start in building up such self-confidence. The various chapters in this book have the same aim: not to tell the teacher what to do but to assist the teacher in choosing between different possible approaches. All over the country are teachers' groups – sometimes linked to Schools Council, Association of Teachers of Mathematics or Mathematical Association – developing the same attitude of self-help. Some of the authors of chapters in this book are concerned in the Diploma in Mathematical Education of the Mathematical Association which aims to raise the level of mathematical self-confidence of primary and middle-school teachers and which includes an individual child study as a central part of the course. Similar help comes from local authority teachers' centres, from the mathematics departments of universities and polytechnics, and from the pages of *Mathematics Teaching* and *Mathematics in School*.

The case made in the Cockcroft Report for more opportunities for in-service training, together with the insistence that such training be school-based, offers the hope that we can build upon the initiatives which have already begun. If in-service training can become an experience of mutual exchange of ideas between responsible teachers at various levels, rather than one-way instruction from the 'expert' with no recent school experience to the 'student' who is exhausted after a heavy day, then the morale and effectiveness of the profession will rapidly improve.

Even in the confusion resulting from falling school rolls, school re-organisation, and economic recession there are are some signs of hope. The Cockcroft Report may help to bring new definitions of the relationship between the compulsory and the optional parts of the mathematics curriculum, between the teacher of mathematics and the teacher of other subjects making use of mathematics, and between provision of mathematics through schools and through further education colleges. From such new definitions, just as from the forthcoming changes in examinations at 16+ and 18+ may come new challenges. In the age when we no longer believe in progress, when each teacher assumes a personal responsibility and when very little help is available from outside 'experts', confusion and challenges are a sign of hope. It is when the individual mathematics teacher is no longer faced with challenges, finds the job easy or accepts too gullibly the guidance of educational administrators or politicians or academic theorists that good mathematics teaching will die.

Bibliography

Hesse, H. (1972) *The Glass Bead Game*, Penguin Books, London.

Skemp, R.R. (1971) *The Psychology of Learning Mathematics*, Penguin Books, London.

Skemp, R.R. (1976) 'Relational Understanding and Instrumental Understanding', *Mathematics Teaching*, *77*, 20–6.

Skemp, R.R. (1979) 'Goals of Learning and Qualities of Understanding', *Mathematics Teaching*, *88*, 44–9.

Tall, D.O. and Vinner, S. (1981) 'Concept Image, Concept Definition in Mathematics with particular reference to limits and continuity', *Educational Studies in Mathematics*, *12*, 151–69.

Wain, G.T. (1978) *Mathematical Education*, Van Nostrand Reinhold, New York and London.

2 TEACHING MATHEMATICS IN PRIMARY AND MIDDLE SCHOOLS

Peter Reynolds

This chapter attempts to review the teaching of mathematics in primary schools in England and Wales. The account can be neither straightforward nor comprehensive since primary schools vary in many ways including size, age range and style of teaching. The responsibility for deciding the curriculum taught in primary schools is largely in the hands of the head-teacher and so, once again, there is wide variation. The main intention of the following paragraphs is to highlight major problems, as seen by a detached but sympathetic observer, with some indication of possible solutions to the difficulties raised.

2.1 A Typical Primary School in the Early 1980s

The description 'primary school' can mean many things from a small one-teacher rural school with under 25 pupils to a very large urban school of nearly 1000 pupils. The age range of such schools can vary widely too. Perhaps the most common pattern involves transfer to a secondary stage at eleven years of age i.e. there are a large number of British primary schools with the age range five to eleven. Even so, there is wide variation, with some schools taking pupils at four years of age, or possibly three years of age where there is appropriate nursery provision. There is also the possibility of a split at seven years of age with the younger children described as 'Infants' and the older pupils as 'Juniors'. Sometimes the split at seven means two quite separate schools under different headteachers.

In those areas where middle schools have been introduced, the primary school is sometimes called a 'first' school. The age range of first schools is usually five to eight or five to nine, which is followed typically by eight to twelve or nine to thirteen middle schools. We will take a closer look at middle schools at the end of this chapter. There are also some five to twelve primary schools.

Apart from the variety of size and age range, primary schools are housed in a wide variety of buildings, from old Victorian buildings, dating back to the mid-nineteenth century, to modern buildings. Some of the more recently-built primary schools are open-plan (i.e. 'classrooms' may not exist in the conventional sense). Most modern classrooms are well provided with facilities for practical work (e.g. sinks for water) but often the size of rooms is such that they are more cramped and with less storage space than in the more spacious older buildings.

The differences between primary schools do not stop at the physical ones already mentioned. Freed from one external pressure by the widespread abolition of the 11+ selection test for grammar school, primary schools were able to experiment with a wide range of styles of teaching. The 1960s witnessed a major change to practical 'discovery' work and 'modern' mathematics, two different movements which made a big impact on the mathematics taught in primary schools. The 1970s saw a retrenchment under increasing external pressure from employers, parents and politicians.

Many of the good things gained in the 1960s were lost in the 1970s. In an attempt to satisfy the often ill-informed public criticism, teachers sometimes returned to the arid mechanical arithmetic of former years. Indeed the HMI survey 'Primary Education in England' (DES, 1978) observed that there was evidence to suggest that some teaching seldom goes beyond repetitive work and results are sometimes disappointing when account is taken of the amount of time that is given to mathematics in primary schools.

The last few years have seen an unprecedented spotlight focused on what we teach in schools. When, at the height of the Great Debate in October 1976, the Prime Minister Mr Callaghan spoke at Ruskin College, his call for an inquiry into the apparent deficiency in basic skills among school leavers was received sympathetically by many people outside education. The Cockcroft Committee of Inquiry was a direct result of Mr Callaghan's plea. The simplistic view taken by the general public was that schools had failed to teach 'basics', such as 'the tables', and schools should pull up their socks. As the Cockcroft Report shows, the teaching of mathematics in primary schools is not quite so simple. Despite this, and other evidence, most adults have a somewhat distorted view of what schools do, and what they should do, based on their own schooling 25 years ago (in the case of parents), or 50 years ago (in the case of many employers). The overall effect of this public scrutiny has been to lower morale amongst teachers and, in particular, to restrict the diet offered in primary schools to those repetitive skills which, although valuable, are of little use without a wider understanding based on practical experience.

The net effect was that a typical primary school by the early 1980s spent an increased time on mathematics, the content of which was often restricted to little more than practising arithmetical skills. A particularly sad feature is that the practical work, so enthusiastically embraced in the 1960s, has disappeared for most pupils by the age of eight. Quite wrongly, many teachers have been convinced that practical work is 'babyish' and 'a waste of time'. So we see primary children getting on with 'serious' work in mathematics without perhaps much idea of what the symbols mean. Another feature of primary schools by 1980 was the widespread reliance on a single published text-book (or work-card) scheme for the main course. Very rarely did one see a school which had devised its own scheme of work. The main scheme was often supplemented with extra drill/practice work. There are many explanations which can be provided for this apparently unsatisfactory state of affairs. An important aspect, which we return to later, is the lack of appropriate training for primary teachers to become leaders for mathematics, especially important since a large number of primary teachers are poorly qualified and insecure in mathematics. Another reason is the way the educational pendulum swings from fashion to fashion, rarely resting in a position of sensible balance. By 1980, the pendulum was firmly in the area of testing and accountability, and yet, in 1970, such thoughts were rarely entertained.

But despite the confusion and lack of clear direction, it would be wrong to assume that disaster has struck. Primary school teachers are dedicated and caring people who have achieved satisfactory results. The teaching of mathematics has not been neglected but it has often lacked clear guidance. On the other hand, with so many changes in the 1970s alone, clear guidance, e.g. in the form of LEA guidelines, has been difficult to provide. Decimal currency, metric weights and measures, calculators and computers have all had, or will have, an immense impact which it is not always easy to anticipate. Nevertheless, every primary school has a duty to produce a scheme of work to guide its teachers.

2.2 Schemes of Work

One of the most revealing aspects of my present work as an LEA adviser is talking to probationary teachers in their first year of teaching. Most young teachers leave college with a streak of idealism and a passionate desire to do the best possible job for the children in their care. All too often, teachers in their first year in primary schools find themselves without an adequate mathematics scheme to follow.

Of course, I have to be aware that it is safer to feign ignorance and an insecure beginner may not wish to admit a lack of grasp of the school's scheme. But I am also aware that often the school's scheme of work has not been given a recent airing. It is all too easy to do a thorough revision of the scheme when a major event takes place, such as the introduction of decimal currency in 1971, and then forget about it for ten years apart from minor alterations. To some extent, in a small primary school, that is satisfactory since all the staff (say five or six teachers) meet frequently and develop a mutual understanding of what changes have developed over the years. It is the newcomer, often in the form of a probationer, who must try to cope with the inadequate written scheme and who quite often becomes distressingly lost.

As we have already noted, a large number of primary schools rely entirely on a published scheme such as Fletcher's *Mathematics for Schools* or *SMP 7–13*, in which case the school's scheme may be no more than a phasing of the material over the age range in the school. Such action almost amounts to dereliction of duty. One can appreciate how it happens since many teachers feel so lacking in confidence that they would never consider 'challenging' the experts who have written the published material. (If only they knew the insecurity of the authors!). Whilst it is unlikely that a particular text-book (published at a particular time, tested with particular pupils and designed for a particular purpose) will be equally appropriate for all pupils in all schools, that is what some publishers claim. It is also what many teachers believe.

It would be foolish to reject all published schemes because none of them fit a particular school totally. Each primary school must be aware that the extremes of ability, say the top 10 per cent and the bottom 10 per cent, will need special treatment – and that is not easy in a mixed-ability class of 30 pupils. We consider different styles of teaching later but it is important to note at this stage that a common but unsatisfactory solution to the problem is that *every* child works at his own pace all the time from published materials not designed to be used in that way. Whatever the publishers claim, it is clearly unrealistic to expect children to teach themselves a difficult subject – teachers must find a method of presenting mathematics in an acceptable form to pupils of all abilities. It is not possible that a published scheme will do that on its own.

Hence we should look more widely for guidance and inspiration as to how to make the best use of a school's resources of staff, rooms, equipment and for the specific needs of the pupils in that school. One particular solution has been the production of *Guidelines* by local education authorities. Typically, these have been produced by a working party of teachers, advisers, Teachers' Centre wardens and others, resulting in a printed booklet arriving in every primary school in the area. There is no doubt about the immense value that this activity has had for the members of the working party, but the receiving schools have often found themselves in the position of the new probationer who was not privy to the discussion and thus did not know the nuances.

Printed guidelines often contain black and white statements. In reality, the decision to include or exclude a particular topic may have been borderline but it appears in the same black type. It is rare for a guideline to give the flavour of the discussion that led up to its production. How can (say) 100 hours of discussion be transmitted in a document that must be understood in one hour? A printed scheme may be no more than a list of skills and knowledge to be achieved on transfer to the next school (at the age of eight, nine, ten, eleven or twelve) thus lacking any indication of emphasis, style of presentation, the practical work required, the pace of development or the relative importance of its components. It is interesting to observe the differences between guidelines produced by different LEAs at different times. In the 1960s the major thrust was to help teachers to incorporate 'modern' mathematics into their teaching. There was an excitement with new approaches, especially the work of the Nuffield Mathematics Project and HMI Miss Edith Biggs, which became incorporated in guidelines and thus provided information about ways of teaching (say) sets or graphical representation. Some of the difficulties mentioned earlier arose because the inevitable shorthand used by the guideline writers could, so easily, be misinterpreted as abandoning all repetitive skills practice.

By the late 1970s, the newly produced guidelines were concerned with assessment, with a feeling that the party was over and a return to sanity was overdue. And so we can see, through guidelines, the swing of the pendulum which has been so harmful. Many voices have been raised in support of the idea that LEAs should put their houses in order and tell schools what to do – but what they might tell them in 1965 may have changed by 1975. We resist the temptation to become entangled in the arguments about whether a centrally-imposed curriculum (by the DES?) would be the best way to proceed.

The reality, however, for most schools, is that each school must decide for itself what to teach, when to teach it, to whom to teach it and how to teach it. There are many ways that this task can be approached. In the absence of LEA guidelines, reference books such as the HMI series handbook *Mathematics 5–11* (DES, 1979) can provide excellent assistance to primary teachers, especially those whose pupils transfer at eleven. Another, perhaps more effective, solution is that every primary school has an adequately-trained teacher who is capable of devising an appropriate scheme of work. This is the major aim of a highly-successful in-service Diploma Course, validated by the Mathematical Association, available at over 50 colleges throughout the country. We return to the development of this Diploma later when we consider the role of co-ordinators and how the Diploma Course prepares teachers for the task.

It is perhaps worth adding a note about LEA advisory services. Although most LEAs now have a mathematics adviser it is likely that he will have between 300 and 500 schools in his mathematical care. It is possible that the LEA also has a primary adviser who can offer practical help and some LEAs have a team of advisory teachers, but the general position is bleak. The average benefit that each primary school obtains from the advisory service must necessarily be superficial, although there are ways, for example by meetings at Teachers' Centres or in groups of schools, in which meagre resources can be made to go a little further.

2.3 Teaching Styles

Traditionally, teachers have been kings in their own castles where they considered what went on as mainly their own affair. But there has been some movement away from this traditional view including architectural imposition (although one does see examples of schools where the open-plan intentions of the planners has been amended by rows of filing cabinets and bookcases). Another factor has been the widespread introduction of work-card schemes which require that pupils work either individually or in small groups. Occasionally we find examples of team teaching where two or more classes are combined for mathematics and their teachers operate as a team, sometimes talking to the whole group, sometimes assisting children individually. Once again we see polarisation. Teachers tend to adopt either the whole class approach or the individualised learning approach in which each pupil theoretically works at his optimum rate on his own. It is perhaps helpful to characterise each of these extremes.

The whole class approach is typified by the teacher providing the initial instruction, using a blackboard and other appropriate aids, with every child in the class performing follow-up work, often in the form of exercises. The merit of this approach is that all pupils are taught simultaneously and all receive the benefit of a few minutes of teacher-instruction. The children hear language used appropriately, they see good layout on the blackboard and generally a high standard is set. The weakness of this approach is that the teacher must aim the level of presentation somewhere in the middle of the ability range which often means that the bright pupils are held back (and consequently often bored) and the slower pupils are taken at a pace which is too fast for their comprehension.

The individualised learning approach has many attractions. It provides an answer to the problem of the bright and slow pupils: every pupil works at his own pace. The form of individualised learning most commonly found is for a class to be working through a series of work-cards with the teacher acting as a low-level supervisor. Mathematics is a difficult subject to teach and even the best work-cards have limitations. It is nearly impossible for the teacher to see every child during each lesson; it is even more difficult to record progress and plan a satisfactory programme for each child. What often happens is that pupils pace themselves (slow or fast according to inclination rather than ability) and the teacher sits at his desk with a queue of children with queries to be answered. The pupils rarely hear the language of mathematics used correctly and errors picked up through self-instruction may remain undetected for some days. It is important to observe that this method requires some competence in reading and this is not appropriate before the age of six or seven.

The two extremes of whole-class and individualised learning are useful poles for our discussion. Recent research by the ORACLE project (Galton and Simon, 1980) makes it clear that a mixture of approaches is likely to be more effective than either extreme separately. Whole-class teaching increases the contact between teacher and pupil, although the extent of the involvement depends on the teacher's skill in probing, querying and general stimulation. Of course the class size is an important factor too: a class discussion with 17 pupils has a different character from one with a class of twice that size. Galton and Simon pointed out that the skills of mixed-ability teaching and the management of the whole class as a unit should be a component of every primary teacher's training. They made the point strongly since their evidence suggested that some of the recent products of colleges lack these skills, and indeed were perhaps never introduced to them. They also observe that unless class size can be reduced to an average of 20, total individualisation must be ruled out as an option. A mixture of the two extremes, together with some group work, appears to be a successful combination.

Group work is an aspect of teaching that many teachers find difficult. It is characterised by a task, such as a practical activity, where a group of two or three pupils, possibly up to six in some cases, work together. It is then possible for the teacher to instruct the group more effectively than individually, and there is the additional benefit of group discovery and discussion.

There are two distinct kinds of practical work, both very desirable, which are ideal for group work. The first is the use of structural apparatus, such as cuisenaire rods and Dienes' MAB, which by analogy helps children to improve their conceptual understanding of the number system. The use of such apparatus should be continued as long as pupils need concrete material to assist their understanding, which for the less able means throughout the primary years. The second kind of practical work concerns the direct experience of weights and measures. Too many pupils are able to perform calculations using metres or kilograms without any feeling for the size of the units. For much measurement work several children are required, for holding tape measures and recording results, and for the subsequent discussion.

As we observed earlier, far too many pupils over the age of eight do no practical work of any kind. Schools defend this stance with reasons such as 'it's messy', 'inconvenient', 'not serious maths' or 'we haven't got the time'. The sad fact is that many children, by the time they leave the primary school, do not understand the mathematics which they have been performing with some competence. Practical work is the only way that they will begin to understand what they have been doing, and some time must be found for it.

2.4 The Role of the Co-ordinator

It is clear that teaching mathematics in primary schools is far from simple. The needs of society change as do the pressures on the educational system. The simple agricultural society of 250 years ago did not require much mathematical ability from the 95 per cent of the population for whom back-breaking work on the land was their life sentence. Today, with automation and the widespread use of computers, less than 5 per cent of the population is involved in agriculture. To cope with the complexity of modern society, it is necessary to know a lot more mathematics than formerly, although for our least-able pupils (say the least able 10 per cent of the population) we may not succeed – they may have to cope with life with little more than a hazy idea of what (for example) income tax at 30 per cent means. Thus, whilst the most able eleven-year-old will certainly be master of most of the arithmetic he will need for life, the least able-pupil will trail a long way behind.

To some extent this may not matter, for most routine jobs in employment have been 'de-skilled' i.e. the employee merely follows a set of rules which lead to the correct answer without the operator needing to know how it works. But education is concerned with much more than a preparation for a job and the sound foundations, which should be provided in the primary years, will help the adult citizen to budget sensibly and to have a better understanding of the society in which he finds himself. We need to know more mathematics today, and different mathematics, than in earlier times.

Although we have mentioned the changes that have taken place over a period of 250 years, quite substantial changes have happened in the short period of 15 years from 1965 to 1980. Major influences on what we teach in primary schools have included the decimalisation of the currency, the introduction of metric weights and measures, the widespread availability of cheap and reliable hand calculators and the development of micro-computers at a price that primary schools can afford. Perhaps we ought to add 'modern mathematics' too, although the new topics, introduced in the 1960s now seem far away. Alongside the new topics, new styles of teaching, such as the use of work-cards, have made the issue more complex.

How can the average primary school teacher, who typically teaches all subjects, keep up-to-date in mathematics? One possible solution, and an increasingly common one, is to give one teacher the responsibility for keeping himself and his colleagues informed. We shall describe such a person as the co-ordinator for mathematics. Such a person would be broadly in charge of all the mathematics taught in the school, although naturally this would necessitate wide co-operation and consultation with others including the head, colleagues teaching in the same school and appropriate neighbouring schools, and LEA advisers.

Perhaps the main duty of a co-ordinator is to ensure that there is an adequate scheme of work, clearly laid out and structured in such a way that all teachers know what is expected of them. This is particularly important in the case of probationary teachers. However it is not enough to have a scheme without ensuring that it is implemented. This implies that there is an adequate stock of resources, including text books, apparatus and other materials, easily available to all teachers as required by the scheme, and that those teachers actually know where to obtain supplies. No scheme of work should last for long without revision. Hence there is a need for the co-ordinator to arrange meetings to discuss the scheme and to enable colleagues to teach more effectively. The co-ordinator must be aware of new developments and present them in simplified form so that an informed discussion can take place. The monitoring of each individual pupil's progress through the scheme is also the responsibility of the co-ordinator. He should assist in the diagnosis of particular strengths and weaknesses and offer advice on appropriate material to stretch the bright pupil or assist the weak pupil.

By appointing a teacher to the post of co-ordinator, a primary school should be able to ensure that mathematics will be given its fair share of display area, suitable library books and general prominence in a way that will create a mathematical environment.

But however lavish the material provision, the co-ordinator will fail to do his job effectively if he fails as a diplomat. Mathematics is a very sensitive area, and personal qualities of a high order are required too. It would also be a tragedy if the appointment of a co-ordinator meant that other teachers gave up their own enthusiasm in the subject. The skills required by co-ordinators are not easily acquired and so, in 1978, the Mathematical Association set up a Diploma in Mathematical Education with the specific intention of preparing suitably-qualified teachers for taking up leadership roles in primary and middle schools.

2.5 The Mathematical Association Diploma

Traditionally, the Mathematical Association, founded in 1871, has been concerned mainly with the teaching of mathematics to secondary pupils. Its first diploma examination paper, set in 1961, was designed to improve the mathematical knowledge of secondary teachers, a mantle later assumed by the Open University. For the new diploma, it was decided that a national examination paper was inappropriate and that each college course should reflect the local needs within a national framework. Typically a college (whose normal work is teacher education) prepares a course proposal which is then submitted to the Mathematical Association's Diploma Board. If the proposal complies with certain guidelines laid down by the Board and provided that a team of visitors from the Board is satisfied with the resources available, then approval is given to run a course. (Approval is also required from the Department of Education and Science.)

The structure of Diploma courses varies, but a common pattern is that the teachers on the course attend a three-hour session at the college, once a week, usually after school has finished, for two years. Together with occasional whole day meetings, the 'contact' time with lecturers must be at least 200 hours. The size of teaching groups is usually in the range 12 to 20 teachers, sufficiently large to enable profitable group participation, sufficiently small not to overload the college's provision of staff, library and other resources.

It is expected that the 200 hours contact time will be divided equally between mathematics and mathematical education, although in the best courses, the integration of the two branches is so effective that it is difficult to separate them. For the award of a Diploma, candidates must satisfy the examiners in each of four components:

(1) mathematics,
(2) a mathematical investigation,
(3) mathematical education,
(4) a special study of a child (or children).

As the Diploma is designed to enable suitably-experienced teachers to become co-ordinators in all primary and middle schools, courses leading to the Diploma attempt to cover the whole age range 5–13, and several of the components above are identical for infants teachers and middle-school teachers. Some aspects of mathematical education give an opportunity to reflect an interest in the more limited age range taught by course members.

Taking the four components in turn, the assessment of mathematics is the only one which the Board insists should be tested mainly by written examination. Although the intention, to ensure that all teachers gaining a Diploma are seen to be competent in mathematics, is desirable, it has caused some unfortunate cases of bad practice. In some cases, teachers have learnt, without understanding (and possibly the night before the written paper) certain tricks and skills. Although the Diploma Board is distressed to hear of such unfortunate examples, generally this component has worked fairly satisfactorily; it could work a lot better if colleges showed more imagination in the types of written papers set, which the guidelines permit, including for example untimed and 'open book' papers.

The second component, a mathematical investigation, is a less traditional way to test mathematics and some people claim that it should be the only way. The investigation, provided by the lecturer or the course member himself, is designed to enable the teacher to learn mathematics by doing it. The topic for investigation may vary from a simple number pattern, rich in opportunities for exploration and conjecture, to quite difficult 'open' situations.

In both the assessment of mathematics and of the mathematical investigation, difficulties arise because of the different backgrounds of the teachers on courses. It has been suggested that it would be more effective if the Diploma recorded the *progress* made by teachers during the course. This might include content and the crucial approach and attitude to mathematics. It is more helpful to know how mathematicians think than to solve examination questions about equivalence relations, but not so easy to organise!

The third component, mathematical education, includes a wide variety of activities, including essays which enable the teacher to read, and observe in the classroom over a period of several weeks, a topic which the normal rush of teaching does not allow. One of the dangers in mathematical education, which has developed a very substantial literature of its own, is that examination paper questions can be set – and answered – without direct reference to the classroom. The Diploma Board has seen examples of examination questions which would have been far better tackled as an essay or some other form of course work assignment.

The fourth, and complementary, component is a special study, in some depth, of a particular child or group of children. Again this forces the busy primary/middle school teacher to stop and reflect on the reasons for success/failure of particular pupils or groups of pupils. It is also an opportunity to teach, in an innovative way or an unfamiliar topic, that normal timetables do not permit.

The first courses leading to the Diploma started in 1978. By 1980, when the first wave of 400 diplomates qualified, it was possible to make some judgement as to whether it had achieved its main objective of helping teachers to assume leadership roles. Throughout every

Diploma course, the role of the external moderator is vital: at the end of the course, the moderator's report provides a fascinating insight into what has taken place.

It is worth quoting two sections from different reports, one very encouraging, one raising doubts:

> The course was characterised by its professional relevances; all of its members spoke highly of its influence upon their classroom teaching and the contribution it had helped them to make towards curriculum development in their schools.

> It must be assumed that these difficulties (the student's difficulties in learning and teaching mathematics) continue to exist even within a course created to try to alleviate them. It is not enough simply to 'try again' to teach them the same mathematics.

We know that, despite some shortcomings, most of the Diploma courses have achieved what they set out to do, that is, to produce better-educated teachers of mathematics who have the confidence and competence to question accepted practice, and who have the skill and diplomacy to carry out what needs to be done.

The Open University has also made its first attempt to produce a course in mathematical education. Under the title 'Mathematics across the Curriculum' (PME 233) (OU, 1980), it is intended to provide an introduction to the subject which will be relevant to the work of primary and secondary teachers, specialist mathematicians and non-mathematicians. The course concentrates rather more on classroom case studies, presented through television film, than the Mathematical Association Diploma course. There is also a greater emphasis on the potency of mathematical ideas, particularly the power of mathematics to explain. Originally PME 233 was designed as a free-standing unit, but it will be possible in future to incorporate it as a half credit counting towards a degree, thus enhancing its attractiveness to teachers. (The Mathematical Association Diploma entitles holders to receive a merit addition to salary.)

2.6 Assessment in the Primary School

What is taught in primary schools, and what others expect of primary schools, is largely determined by tradition. Teachers often teach in the way in which they were taught. Parents expect their children to do what they themselves did a generation earlier. Secondary schools expect pupils to be ready to start Book 1 of the chosen text-book when they transfer. But it is only in the last few years that it has been possible to determine realistic objectives for primary schools. The recent reports of the Assessment of Performance Unit (APU) (DES, 1980) have enabled teachers to realise that the traditional content, often determined by the 11+ transfer test, was realistic only for the top 25 per cent of the ability range. A moment's thought will tell us that, at the age of eleven, a typical mixed ability class of pupils will contain a range of IQs from 70 to 130. This means that the 'mathematical age' of such a class is likely to range from less than eight years of age to over 14 years of age i.e. from still needing concrete materials through to those capable of abstract thought. Clearly, with such a wide range it is foolish to state even a simple list of objectives which *all* pupils must attain by the end of the primary years. And yet many LEA guidelines and school schemes of work appear to do just that.

The APU Primary Survey Report No. 1 can be criticised in a variety of ways. Perhaps the results would have been different if the questions had been posed in a slightly different manner, perhaps the attempt to establish a national standard is dubious, but we cannot deny that if only 29 per cent of eleven-year olds can calculate 314 × 201 correctly, then it does cast some doubt on the wisdom of teaching long multiplication to all primary children. It may surprise many people that only 54 per cent of the eleven-year olds can correctly put in order of size the numbers 275, 752, 725, 572, despite the efforts made to encourage work with structural apparatus which assists the understanding of place value.

It is easy to continue such a list of apparently dismal results. But we delude ourselves if we think that more efficient teaching will improve the situation substantially. As the evidence continues to grow it becomes clearer that for many pupils we expect too much. Some fascinating work done at Chelsea College, London, for the project Concepts in Secondary Mathematics and Science (CSMS) (Hart, 1981) has made it abundantly clear that too high expectations continue into the secondary level. We may reasonably expect nearly all eleven-year-old pupils to know that 3 × 6 = 18 but many of them will not understand the *situations* in which it is appropriate to multiply 3 × 6 for a further two years or more. Learning mathematics is a very slow and painful business for many pupils, and it is made more painful by attempts to teach to all primary pupils difficult techniques such as long multiplication and long division which might be better delayed for a year or two – or tackled by calculator.

A less-publicised feature of the APU Survey was the practical test in which trained testers interviewed pupils individually. One tester wrote: 'It's useless my child getting all his sums right if he hasn't a clue when you give him some simple apparatus'. Another tester wrote: 'Having seen some of the children floundering, I am now more than ever convinced that before you can understand in the concrete you've got to spend a lot of time working with abstract material'.

Do we feel pleased that 84 per cent of 11-year olds can measure a 13 cm straight line accurately and unaided, or are we concerned that 14 per cent failed even with help? Are we surprised that when given three red and three yellow square plastic tiles and asked 'what fraction of these squares is red?' only 61 per cent gave correct answers even when $3/6$ was allowed?

The problems posed by these questions have been exacerbated by the post Great Debate jumpiness of teachers. The public criticism about poor standards has often been misplaced but has had the appearance of reason. Teachers are now in a better position to state that it is unreasonable to expect *all* children to achieve a certain standard by the age of eleven, or any other age.

The monitoring of pupils' progress is an essential feature of good teaching. Teachers must know whether what they have taught has been learnt. They must diagnose weakness and design suitable remedial work. Able pupils must be stretched. However, it is of paramount importance that the standards demanded by teachers are realistic. Mathematics is not an easy subject to teach, and the success of the teacher depends on setting a sufficiently difficult target for every pupil without demanding too much. On one side of this dividing line lies boredom and lack of incentive, on the other side lies a brick wall of impossible tasks and depressing failure. Assessment must be the servant of the teacher, not the whip of an unsympathetic public which drives teachers to take the road of failure for many children.

2.7 Modern Technology and its Implications for Primary Schools

Recent years have witnessed an ever-increasing range of new products which have serious implications for the primary classroom. Some of these are dealt with in Chapter 7 but we wish to discuss here the overall effect of new technologies on the teaching of mathematics. Primary schools already make substantial use of the excellent television programmes which are designed to assist learning. Many primary schools now have a videotape recorder and thus more flexible use can be made of available material to supplement normal teaching. Perhaps the most useful way in which television can be used with younger children is to stimulate interest which is then developed by the teacher. It does not seem likely that television will replace day-to-day teaching. The imminent availability of cheap video discs will enhance the possibilities, but it will require cheaply available interactive programmes, in which the instruction to the pupil is modified according to his previous answer, before a change of style is possible.

Such interactive dialogue is not far off. British Telecom's PRESTEL service, in which the user receives on his television screen, information via a telephone wire from a central computer, may be able to offer a limited additional service soon. The difference between modern developments, using a computer, and earlier forms of programmed learning is the immense flexibility provided by the computer. Not only will the programme vary according to the needs of the user but modern computers can provide animated diagrams, photographs and other visual aids not possible with earlier technology.

The effect of the computer in primary classrooms is likely to be most noticeable through a television set, whether as a receiver for a central service or as the display unit of a small micro-computer in the classroom. The effect of the computer is also more likely to be as a visual aid or as a teaching device, than as a tool. That is, it is more likely that children will use a computer for which programs have been written to provide 'tables practice' or to teach (say) about fractions. It is unlikely that many primary pupils or their teachers will become skilful enough to write efficient programs, although, clearly, some simple programming will enable users to understand better what is happening and, possibly, to amend a program for the school's special needs.

Whilst the cost of some of the possibilities mentioned so far is still prohibitive for some small schools, the cost of calculators has continued to fall. It is now common for six-year-olds to have a calculator as a Christmas present only to find that he cannot use it at school! Technology moves too fast for the teaching profession, for whom ten years is a relatively short span. It takes ten years for the implications of a change in education to be assessed: in technology a new development appears every six months. There is a serious mismatch which, during recent years, has made teachers appear slow-witted and stupidly conservative.

Teachers have developed skills for teaching mathematics effectively. To make sudden changes, however attractive, causes serious problems. It is now possible to abandon the teaching of long multiplication and long division since those who require those skills can learn to use a calculator. Equally there is now a real possibility that young children will meet negative numbers and square roots long before the age at which they met them in the traditional syllabus. The alarm that this causes teachers is based on several points. First, there is the genuine concern that the secondary school (or parents or employers) may unwittingly and harmfully be affected by the change. Secondly, and linked with the first reason, it takes many years before the effects, good or bad, are noticed. Thirdly, there is a lack of knowledge as to the best way to teach the use of calculators in primary schools. On this last point, it will take some years before teachers are thoroughly conversant with the most effective methods for teaching the use of calculators. We can be certain that teachers will need to spend a lot of time on the size of the answer, approximation, significant figures, rounding errors and the many other difficulties which arise as soon as a calculator is used. We do not need an APU survey to tell us that some children will find many of these ideas difficult! What we can be certain of is that the reluctant acceptance of calculators in the classroom, possibly only for the occasional checking of answers, is no longer acceptable.

Lurking behind any discussion on the use of technology in schools is the question of whether teachers might be replaced by machines. There is now a general acceptance that teachers cannot be replaced. Modern technology enables the teacher to do more teaching and to be more effective, but the teacher is still required as a manager of the available resources, although there are some cases where 'distance learning', as exemplified by the Open University's television, radio and written material by post, works quite well. It is also true that pupils feel less threatened by a machine, which provides a 'tables test', than they would by a teacher doing the same thing. Hence schools need to be aware of ways in which technology does a better job than a teacher, and to ensure that the better job is done.

2.8 Middle Schools

Middle schools do not fit easily into the categories normally used. When some local education authorities 'went comprehensive' it was decided to have a three-tier system with middle schools, commonly 8–12 or

9–13. The advantage of middle schools are that they are smaller and thus more intimate than large secondary schools, and, in rural areas, the travelling distances of pupils are reduced. A major disadvantage is that the small size does not make an attractive career for teachers of mathematics since it is rare for the co-ordinator of mathematics in a middle school to be on a Scale 3 salary for mathematics alone.

There is common mythology that 8–12 schools are more like primary schools and 9–13 schools are more like secondary schools (which they are deemed to be by the DES). The truth is that they are all different, often reflecting the background of the teachers appointed when the middle school was opened. Very few middle schools, however, have a coherent mathematics programme for the whole age span. The traditional British break in schooling dies hard and even in good middle schools we see a marked change as pupils move past the magical age of eleven. To generalise, at the risk of caricature, many middle schools operate as primary schools up to the age of eleven, and as secondary schools after eleven. The secondary school which receives pupils from middle schools at twelve or thirteen often demands, although some only request, that pupils start the O-level text-book series used in the upper/high school since most O-level series are designed as a five year 11–16 course. Thus, since not all pupils are suited to an O-level course, middle schools find themselves forced into setting and many other ways of operating which are commonly found in 11–16 or 11–18 schools.

Apart from this pressure, there is the commercial fact that very few middle school materials are published because of the relatively small market. Even schemes which span the age range, such as SMP 7–13 work-cards and Fletcher's *Mathematics for Schools* are rarely used beyond the age of eleven. And so we often see a split, where children over eleven are taught by specialist teachers (if the school can find enough), usually in sets, and younger pupils are taught by a class teacher who teaches all subjects. The vital task of the mathematics co-ordinator in a middle school is not easy. He will find it very difficult actually to teach pupils in each of the four year-groups and thus have direct contact with the full age and ability range. More often than not, he will be a specialist teacher for three or four classes of older pupils, which together with other lessons (often in other subjects) and other duties, will leave very little time, even if the timetable permits, to see the remainder of the 16 classes which make up a typical middle school.

Mathematical liaison with other schools is always important; it can be an especially thorny problem for middle schools. At its worst, it may mean establishing a contact with a dozen or more feeder primary schools and as many upper schools in those LEAs where geography and parental choice enable a wide choice to be made. Even at its best, in a tightly-drawn rural catchment area, where it is almost impossible for pupils to go elsewhere, there are still severe problems which are rarely solved completely. The essential aim of all liaison between schools is to ensure that there is continuity of content, notation and layout as a pupil moves from school to school. Despite meetings of teachers and visits, misunderstandings can and do occur. It is not easy to specify exactly what skills and knowledge should be achieved on transfer, either into, or out of, the middle school. The issue is further complicated since the full range of ability means that the brightest pupil on *entering* the middle school knows more mathematics than the dullest pupil on leaving!

Middle schools can be very pleasant and exciting places, for pupils and teachers. Pupils tend to be enthusiastic, ranging from the immature child barely out of the infants classroom to the most able 13-year old who is well able to cope with abstract ideas. Given well-qualified teachers and a lively co-ordinator, there is no better placed for pupils in the middle years to be educated.

Bibliography

DES (1978) *Primary Education in England*, HMSO, London.
DES (1979) *Mathematics 5–11*, HMSO, London.
DES (1980) *Mathematical Development. Primary Survey. Report No. 1*, HMSO, London.
Galton, M. and Simon, B. (1980) *Progress and Performance in the Primary Classroom*, Routledge and Kegan Paul, London.
Hart, K.M. (1981) *Children's Understanding of Mathematics: 11–16*, John Murray, London.
OU (1980) *Mathematics Across the Curriculum*, Open University Course PME 233.

3 TEACHING MATHEMATICS IN SECONDARY SCHOOLS

Michael Cornelius

3.1 Secondary School Mathematics Today

Secondary school mathematics is in a mess. Many teachers and pupils are in a state of bewilderment and confusion. There are some teachers who have managed to maintain in their schools a good, balanced mathematics programme and who have the ability to motivate and interest their pupils, but in a large number of schools inadequately qualified teachers flounder as they receive conflicting advice from different directions. There are, unfortunately, too many poor teachers who (perhaps because they teach mathematics through no choice of their own anyway) fail to inspire or excite and reduce mathematics to a monotonous piece of drudgery.

Until about 1960, grammar schools taught a diet of formal mathematics leading pupils to external examinations and concentrating largely on mastery of skills and techniques: no-one seemed unduly concerned about interest, enjoyment or relevance. In the secondary-modern schools few seemed to care about what went on: mathematics was not a subject to be taken seriously by the less able pupil and there were no public examinations to worry about anyway. The next decade witnessed one of the most dramatic and traumatic upheavals which secondary school mathematics has ever experienced. First came developments in the content, teaching and presentation of mathematics and the birth of many 'projects' and secondly, just as many teachers were facing what they saw as a dilemma in choosing between 'modern' and 'traditional' mathematics, the selection of pupils at the age of eleven was largely abolished and those who had previously taught only groups of fairly homogeneous ability had to face the new challenge of being comprehensive school teachers and the problem of wide ranges of pupil ability. During the 1960s 'discovery learning', 'concept acquisition' and 'learning by doing' became part of the fashionable mathematics teacher's vocabulary and basic skills and techniques were often relegated to a lowly position as relevance and involvement became key goals. Content was turned upside down, formal geometry was an early casualty, and attractively-presented but often superficial new areas like sets, number bases, topology or matrices replaced much of the former manipulative algebra or arithmetic. New projects and courses were eagerly taken up and one of the leading pioneers, the School Mathematics Project (SMP), for example, showed an increase in O-level entries for its examination from 919 candidates in 1964 to 20,100 in 1970 with undoubtedly a large number of others using the project materials (Thwaites, 1972). There were many who jumped quickly onto the fast-moving bandwagon of new mathematics in schools. Few voices were raised in protest although one of the early pleas for caution came from Dr J.M. Hammersley in a lecture to the Institute of Mathematics and its Applications in 1967 'On the enfeeblement of mathematical skills by "Modern Mathematics" and by similar soft intellectual trash in schools and universities' (Hammersley, 1968) in which he argued for moderation in the use of mathematical jargon and abstract mathematical structure in school mathematics, singling out in particular the areas of set theory, definitions and foundations of the real number system, and abstract algebra and vector spaces.

In the early 1970s, as some of the glossy packages of the previous decade began to lose their appeal, teachers were bombarded with advice and suggestions about how they might cope with low-ability pupils, mixed-ability groups, gifted children, work-cards and work-sheets and all this at a time when they were desperately trying to learn to live with the calculator and computer. Concern began to be expressed about standards in school mathematics and about inadequacies in the mathematical backgrounds of school leavers as perceived by employers. More voices, both official and unofficial, began to question the state of school mathematics until finally a Committee of Inquiry into the Teaching of Mathematics in Schools, the Cockcroft Committee, was established, its report eventually appearing in 1982.

Inevitably the key to good, successful learning of mathematics is the *teacher*. The teaching of mathematics in secondary schools has for many years suffered through an acute shortage of qualified teachers. Since the 1950s, when mathematics graduates found attractive opportunities in professional, industrial or academic employment, schools have had to live with an inadequate supply of mathematics teachers. The DES survey of secondary schools (DES, 1980) indicates that of the 3,365 teachers of mathematics included in the survey: 27 per cent were graduates in mathematics; 28 per cent were non-graduates who had studied mathematics as their main subject; 22 per cent, whether graduates or not, had mathematics as their *second* subject of qualification; and 23 per cent did not have mathematics as either their first or second subject of qualification.

As the secondary school population falls during the 1980s, and many schools close or contract in size, there is an extreme danger that vacant posts will be filled by redeployment within local authorities, thus perhaps *increasing* the number of unqualified mathematics teachers and leaving able potential new entrants to the profession unemployed. It is unfortunate that when, for the first time for many years, there is the possibility of improving the standard of secondary mathematics teaching through an adequate supply of teachers, falling school rolls and financial constraints may in fact cause the position to worsen. The Cockcroft Committee has recently commented on this problem.

In view of the change and upheaval which has faced teachers, particularly mathematics teachers, in the last twenty years, it is not surprising that many are overwhelmed, confused and uncertain about what to do or which way to turn. This chapter seeks to examine our reasons for teaching mathematics in secondary schools, to look at some of the problems involved and to try to point the way to a middle road between the old formality of skill acquisition and the more recent emphasis on a freer, less formal and open-ended approach.

3.2 Why Teach Mathematics?

> The great problem of mathematics education is the gap between use and aim; in no other field of instruction is the distance between useless aim and aimless use so great. (Freudenthal, 1973)

Only if some broad measure of agreement can be reached in answer to the question 'Why teach mathematics?' can we begin to discuss *what* we teach and *how* we might plan and organise the teaching. In discussion with a group of secondary school mathematics teachers the following aims were suggested:

(1) to develop the ability to think, communicate and reason clearly and logically,
(2) to provide tools and skills necessary for use in the real world, everyday life and other subjects,
(3) to develop the ability to recognise patterns and relationships and

to generalise from experience (including the use of symbols),
(4) to develop creative ability,
(5) to increase awareness of other cultures and interest in the world.

These rather general aims would probably receive fairly wide acceptance among secondary school mathematics teachers, at least in public. In private many would confess to the more immediately practical aim of helping pupils to pass examinations. Any set of aims is excellent in theory but, in practice, difficult to translate into the day-to-day routine of classroom teaching. Pressures are often such that teachers, whilst paying lip-service to acceptable goals, succumb to the temptation to satisfy what they see as the most necessary, immediate needs. Thus there is likely to be a gap between what teachers *say* and what they *do*. Most would probably like to claim that they teach mathematics which is useful to pupils in everyday life and in other subjects, promotes mathematical awareness and creates interest and enjoyment. Few honest teachers would claim to achieve these aims with anything more than a tiny minority of pupils.

In the DES survey of secondary schools '*Aspects of Secondary Education in England*' (DES, 1980), although no direct discussion of aims is included, the chapter on mathematics includes a list of points which were considered in assessing the *provision* by the school and the *quality of response* by the pupils:

(1) pupils' oral communication with their peers and with their teachers, their ability to understand the printed page and to write clearly for the benefit of others
(2) evidence of understanding provided by appropriate practical performance
(3) evidence of sustained work on any mathematical topics, including the ability to read from topic books or to conduct an extended investigation
(4) evidence of productive work in groups
(5) evidence of profitable links with other areas of the curriculum
(6) whether the pupils were willing to work independently or whether they needed detailed instructions on method every time.

These points would seem to provide a most useful guide for a teacher who is moving from the question 'why?' to the question 'how?' and is thus in the process of translating aims into classroom practice. It is indeed interesting that oral communication appears high on the list:

many secondary mathematics teachers have apparently abandoned spoken mathematics as they feed their pupils on an exclusive diet of individual work-cards to such an extent that we find pupils who, for example, can *write* the symbol 'π' but are unable to read it as 'pi'! The DES survey observes that:

> Very little opportunity was given for the pupils to express themselves orally in mathematics lessons and this resulted in some very confused statements in the exercise books when anything in the nature of an explanation was demanded.

Another piece of work which may shed some light on the translation of aims into classroom practice is the American National Council of Teachers of Mathematics (NCTM) set of recommendations for school mathematics in the 1980s (NCTM, 1981). The first 3 recommendations are: '*Recommendation 1*: Problem solving must be the focus of school mathematics in the 1980s. *Recommendation 2*: The concept of basic skills in mathematics must encompass more than computational facility. *Recommendation 3*: Mathematics programs must take full advantage of the power of calculators and computers at all levels.'

The view of school mathematics as a problem-solving activity is one which seems to be gaining in popularity on both sides of the Atlantic – for example in this country several Open University courses give strong emphasis to this aspect. The NCTM concludes, in a survey which followed up the publication of its recommendations, that the climate for implementing problem solving as a focus is highly favourable but the task of hammering out details is no small one for, despite surface agreement, perceptions of what problem solving could entail differ widely. Likewise, in implementing Recommendation 2, there is a need to agree on what encompasses basic skills and what level of attainment should be expected; there are likely to be strongly differing views from teachers and others on this point, particularly at a time when the use of calculators is becoming almost universal. It seems improbable that many people would disagree with Recommendation 3 but here again it is not easy to reach total agreement about calculator issues and the NCTM survey reports sharper differences here than on any other issue, concluding that changing the curriculum to incorporate the use of computers could proceed much more smoothly than changing the curriculum to incorporate the use of calculators. A fuller discussion of the use of calculators in schools is to be found in Chapter 7.

3.3 Mathematics in the Secondary Classroom

Most teachers of secondary mathematics would like an easy answer to the question, 'How can pupils acquire appropriate skills and pass examinations and at the same time be persuaded that mathematics is interesting, enjoyable, relevant and exciting?' Is it possible to satisfy the needs of pupils, parents, schools and society in general and at the same time achieve the undeniably sensible aims of providing interest, enjoyment, relevance and excitement? Under heavy external pressures, most teachers (particularly those inadequately qualified or experienced or those lacking in confidence) concentrate heavily on the acquisition of skills and techniques, with the result that their pupils come to see mathematics as a dull collection of routines which must be learned and mastered. Examinations have to be passed and 'uses' for the mathematics so painfully and laboriously learnt are conveniently ignored or dismissed with a promise that appropriate application will appear one day in the future. The Open University course 'Mathematics across the Curriculum' (Open University, 1980) argues that skill-getting in isolation is unrewarding if every opportunity is not taken for use to be made of that skill and the course team questions whether skills need to be thoroughly acquired before they can be used. This philosophy is illustrated with the diagram:

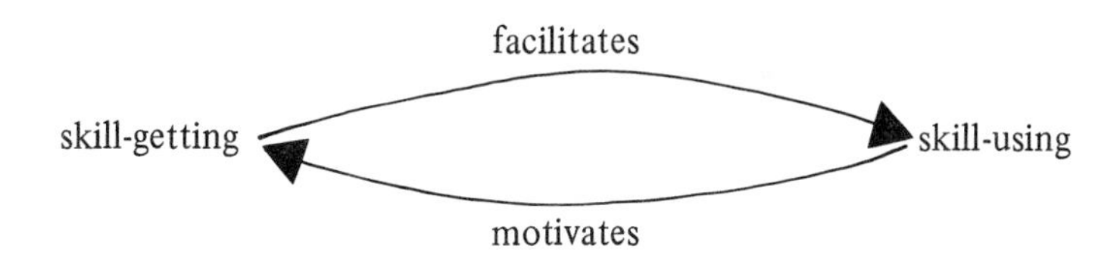

Most teachers would be delighted to have sufficient ideas and time to approach each new topic through a real life or 'skill-using' situation. In practice time is limited and it is not easy to invent or discover enough relevant material. Should a piece of mathematics be taught with a promise that it will be useful later on or should the need for a skill be demonstrated before that skill is introduced? The dilemma for the teacher is summarised in the Open University diagram – does the 'real' problem come first, only to be frustrated in solution by lack of appropriate pupil skills, or are the skills developed first and then applied to a problem? It is not easy for a teacher of mathematics to achieve an ideal compromise. Lessons A and B, reproduced below, are extracts from the notebook of a student on teaching practice (Trepte,

1981) and illustrate some of the difficulties facing not only students but also experienced teachers:

Lesson Plan A
2J: 25 February 1981 10.50 – 12.00
Subject: Solving equations using the distributive law.
Aim: To use the distributive law in solving linear equations and inequations.
(1) Discuss like and unlike terms – giving notes.
Heading: Collecting like terms

Like terms: e.g. $4x, 10x, \frac{3x}{2}$ $3, 5, 10, -5$

$6xy, 2xy, -4yx$
$x^2, -5x^2$
$2x^2y, 3yx^2$

Unlike terms: e.g. $4x, 3y$
$2x^2, 2x$
$3xy, -x^2$
x^2y, xy^2

Like terms can be added together, i.e. $4x^2y + 6x^2y = 10x^2y$
Unlike terms cannot be simplifed, i.e. $4x + 3y = 4x + 3y$

Discuss Ex. 4B (Qu. 1-15)

(2) Solving equations.
Look at $5x + 2x = 14$
Using the distributive law, this can be simplified to

$$7x = 14$$
$$\therefore x = 2$$

Further example

$$6n + 2(n - 2) - 4 = 0$$

Simplify

$$6n + 2n - 4 - 4 = 0$$
$$8n - 8 = 0$$
$$8n = 8$$
$$n = 1$$

When solving equations, keep the equals signs directly below each other

Comments: This lesson went very badly. The class seemed unable to do the questions in the first part of the lesson which meant that the lesson went no further. Unfortunately there was a complete misunderstanding about the questions due to lack of examples, which I thought were unnecessary as they were just the same as the homework. Next lesson must be completely replanned.

Lesson A is typical of many mathematics lessons. Here it was given by a good, conscientious student who was quick to appreciate shortcomings. But this lesson does illustrate, what is widespread in mathematics teaching, a heavy emphasis on skill acquisition. Here a group of 12–13-year-old pupils is being asked to work with abstract concepts in symbolic notation in a situation which has no links whatsoever with anything within their experience. Thus the class finds the work difficult and pupils are 'unable to do the questions'. The likelihood subsequently is that the teacher in such a position will resort to more and more illustrative examples and eventually pupils will do numerous questions in imitative style until they have apparently mastered the technique and can 'do' the questions. Although skills in mathematics are important, the lesson shows how it is easy to become obsessed with technique (and possibly examination needs) and have no hint of immediate relevance, use or interest.

Two questions arise immediately: (a) Why are we teaching this material anyway? and (b) If it has to be taught, is it possible to find a less routine way? The good teacher will have an answer to (a) and will have given much thought to possible answers to (b). The weak teacher will probably not even be aware that these questions might be asked!

Lesson Plan B

30: 20 January 1981 10.00 – 10.35 and 2.35 – 3.10

Subject: Trigonometry – Circumference of a Circle.

Aim: For the class to find the relationship of the circumference and diameter of a circle by practical methods and then to apply the formula.

Class needs: Pencils, compasses, ruler, scrap paper, string.

Introduction: First talk about the importance of the circle but the difficulty of finding its area because it has no straight edges. First look at the circumference.

Development: On the blackboard draw the diagrams that the class are to draw and work with.

On rough paper, get them to draw:

(1)

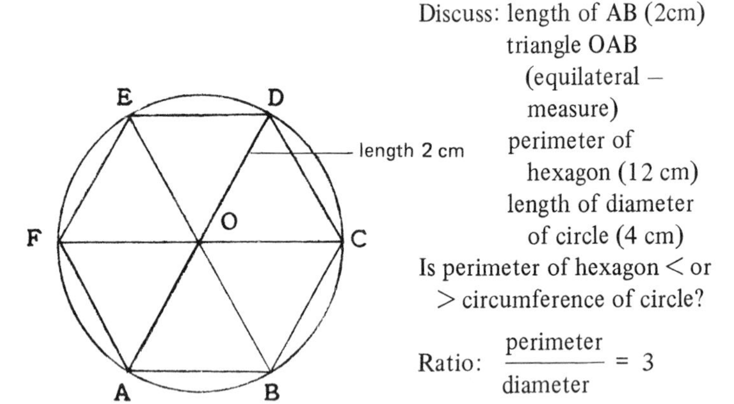

Discuss: length of AB (2cm)
triangle OAB (equilateral – measure)
perimeter of hexagon (12 cm)
length of diameter of circle (4 cm)

Is perimeter of hexagon $<$ or $>$ circumference of circle?

Ratio: $\dfrac{\text{perimeter}}{\text{diameter}} = 3$

So we can say:
circumference is $> 3 \times$ length of diameter.

Draw diagram in exercise books and write down the above facts.

(2) On rough paper again:

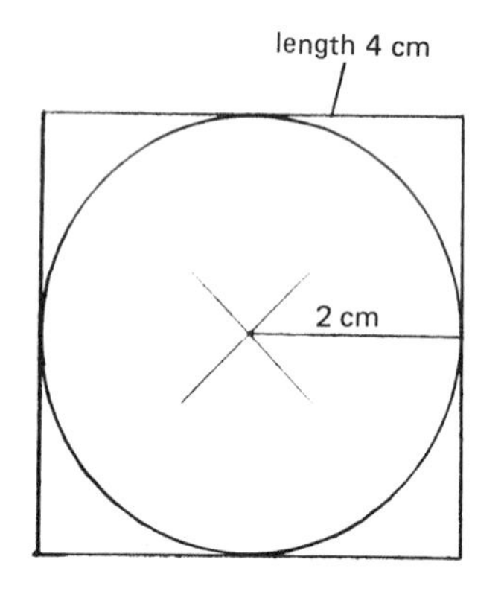

Discuss: length of radius of circle (2 cm)
length of diameter (4 cm)
perimeter of square (16 cm)

Is the perimeter of square $<$ or $>$ circumference of circle?

Ratio: $\dfrac{\text{perimeter}}{\text{diameter}} = 4$

So we can say:
circumference is $< 4 \times$ length of diameter.

Draw diagram in exercise books and write out above facts.

We now know that the circumference of a circle is between 3 and 4 times the diameter.
Will this number vary from circle to circle?
How can we find the number more exactly?

Tell each child to select one coin and one pencil or pen.
Draw 4 circles on rough paper each of a different size with radius less than 4 cm.
One more round object.
They must measure for each object or circle:

Length of diameter
Length of circumference } Accuracy is *very* important!

Draw a table in books:

Object	Diameter (cm)	Circumference (cm)	$\frac{\text{Circumference}}{\text{Diameter}}$

Fill in the table.
What do you notice about the last column?
How do we calculate the average of the ratios in the last column?
Add them all up and divide by the number of ratios.
Then calculate a class average.
What is the average?
(should be approximately π)

Introduce π as a Greek letter p.

$$\frac{C}{d} = \frac{\text{circumference}}{\text{diameter}} = \pi$$

$\times$ top and bottom by d

$C = d\pi$ Now $2r = d$
so $C = 2\pi r$ r = radius

Does anybody know a value for π?
p. 192 3.14(2 (16))

Mention $\frac{22}{7}$ (work out on calculators)

Can use $\frac{22}{7}$ when we only require accuracy to 3 significant figures.

Examples of calculations:
(1) Calculate the circumference of a wheel radius 10 cm

$$C = 2\pi r$$
$$C = 2 \times \pi \times 10$$
$$C = 20 \times 3.14$$
$$C = 62.8 \text{ cm}$$

(2) A wheel has diameter 56 cm. How many times will it turn on a vehicle that travels 440 m?

$$C = \pi d$$
$$= 56 \times \frac{22}{7} = 176 \text{ cm}$$

$$440 \text{ m} = 44{,}000 \text{ cm}$$

$$\text{No. of turns} = \frac{44000}{176} \simeq 250$$

Homework: P. 193, Ex. 2, Qu. 1, 2, 3, 6, 7, 8

Comments: The class were very quiet during both lessons even whilst doing the measuring with the string. They were puzzled at first as to what the lesson was about but once they started getting answers around 3.1 and 3.2 they realised that they were calculating π. Over half the class had heard of it but they were used to the form $^{22}/_{7}$ which there was not time to introduce. The first page of the lesson took longer than expected and so there was no time in the second lesson to go over the second example. It would have been better to leave the multiplication signs in the formulae as they seemed confused about how to put the numbers into them. These were added at the end of the lesson.

After going over the homework in the next lesson, it would be as well to spend time doing some examples of the second type before moving on.

Lesson B includes much more pupil activity and involvement than Lesson A. It is a move away from the straightforward 'technique acquisition' lesson towards an investigative, discovery approach and, whilst it does end with routine calculation, it attempts to provide some scope for pupil initiative and imagination. Doubtless many experienced teachers would have seized the opportunity to introduce some historical background (e.g. Greek 'exhaustion' methods for the measurement of areas, biblical references to the measurement of circles (I Kings 7,23 or II Chronicles 4,2) or perhaps some comments on when the symbol π was first used) and maybe a more established teacher would let the investigation part of the lesson continue longer and worry less about the need to move to the routine solution of questions. But the lesson is basically a sound one for a beginning teacher and makes an interesting contrast with Lesson A. The end products of the two lessons are not dissimilar; in both instances the pupils will have learnt a particular technique, but in Lesson B (unlike Lesson A), there is at least a link with the real world and a sense of some earnest activity on the part of the pupils.

From the reforms of school mathematics in the 1960s came greater emphasis on structure in mathematics and pupil activity (doing, being involved, pleasure and so on). The second of these emphases resulted in much good, purposeful work and a move away from dull routines. Unfortunately the departure from skill acquisition was frequently almost total, with the result that children lacked essential basic techniques and spent too much time simply 'messing around' in an unorganised way. There is, in school mathematics as in most things, a need for a balance between the two extremes; Lesson B above, although it might be criticised as being too formal, is a move towards the point of balance. The attempt to emphasise structure in mathematics at an early stage has led to a large number of total disasters.

The professional mathematician is capable of looking at his subject from a position of both breadth and depth of knowledge and he is able to perceive patterns and structures which interest, excite and fascinate. The immature secondary pupil is unlikely to be able to appreciate these structures in the same way. During the last twenty years work on e.g. number bases, matrices, topology or finite arithmetic has crept into the school syllabus often in a highly artificial way and often with the justification that pupils will be helped to appreciate common structures. It is pertinent to contrast these rather abstract and, at this level, largely useless topics with areas like linear programming, probability and statistics which have also come into school mathematics but which have obvious relevance and possible application. Examinations for both GCE and CSE frequently include questions which encourage the teaching of repetitive routines in areas like matrices, number bases or finite arithmetic. Questions like: Question 1: Multiply 27_8 by 35_8 and express your answer in base 2, or Question 2: If $A = \begin{pmatrix} 1 & 2 \\ 3 & 4 \end{pmatrix}$ and $B = \begin{pmatrix} 4 & 5 \\ 2 & 3 \end{pmatrix}$ find the product AB and the inverse of A are certainly no better than many of the questions they have replaced which encouraged tedious teaching of arithmetic and algebraic technique. Yet there are many teachers who feel that, by teaching number bases or matrices, they are up to date and giving pupils an interesting, exciting modern diet. The research programme *Concepts in Secondary Mathematics and Science* (Hart, 1981; published report titled *Children's Understanding of Mathematics: 11–16*) says of structure and matrices: 'Structure in general appears to be a difficult subject for children and the inclusion of matrices in a syllabus mainly to demonstrate structure is a debatable strategy.'

Undoubtedly generations of pupils were bored with seemingly pointless formal theorems in geometry, with complicated manipulation in algebra and highly artificial arithmetic. It is often argued that the 'I was never any good at mathematics' attitude found among many adults today is a direct result of an unstimulating, uninteresting and largely irrelevant mathematical upbringing. It is probable that the pupil of the 1980s will emerge with the same attitude since, in many schools, the mathematical activity is no better (and indeed may be even worse) than that of twenty or thirty years ago. If it was argued that much of the 'traditional' mathematics was not relevant to a child's needs and experience, then the same accusation can certainly be levelled at many topics in a contemporary school mathematics syllabus. Rarely is the position of mathematics in a school challenged; it is quietly accepted that it is a major subject requiring almost every pupil to endure five lessons per week up to the age of 16 and also being important enough to deserve the status of *two* subjects at Advanced Level. Is it so important? Does *every* pupil need a thousand lessons during his secondary school career? Might not those pupils whose likely destination is not a future career in higher education or science or engineering get by quite happily with a somewhat smaller dose of mathematics? There must be better things to do with a 'fifth form non-examination set' than spending several hours doing routine work on number bases in a dull, mechanical way. Perhaps a rather stronger challenge asking, 'Why teach mathematics?' from teachers in other school subject areas might not only make the mathematician think more carefully about his role but could also lead to a complete reappraisal of the place of mathematics in the secondary school curriculum. For too long the status and quantity of mathematics in schools have gone unchallenged.

3.4 The Organisation of Secondary Mathematics

The move away from selective schools, which accelerated rapidly in the

Table 3.1: Organisation of Pupils, Years 1–3

Year	School A	B	C	D	E
1	All Ability (French setting)	All Ability	3 Ability Bands + Remedial Group	All Ability	2 Ability Bands + Remedial Group
2	All Ability (some setting)	All Ability (some setting)	"	3 Ability Bands + setting	2 Ability Bands + Language Class + Remedial
3	"	"	"	"	"
Average size of year group	150 (6 classes)	240 (8 classes)	120 (5 classes)	210 (7 classes)	200 (7 classes)

late 1960s and early 1970s, often under political pressure, has led to a radical rethinking of the organisation of pupil groupings. Previously most teachers were faced with groups of reasonably homogeneous ability and taught in what is often called a 'traditional' way – all pupils worked on the same topic at the same time and a lesson was normally divided into sessions where the teacher explained or discussed with the whole class and sessions in which pupils worked at examples. The reorganisation of schools led to a situation where the 'traditional' model was often inadequate.

In the lower part of a secondary school the problem of groups of mixed ability is often most acute. Table 3.1 shows the organisation for the first three years in five County Durham comprehensive schools. The particular pupil groups imposed by the schools produce constraints on the teaching of mathematics. The head of mathematics in school D comments: 'I would prefer to set mathematics in years 1–3 across the year group, but this is not possible due to split-site', whilst the head of mathematics in school A comments: 'The school organisation affects mathematics greatly – mixed ability first year determines syllabus – groupings in years 2 and 3 reduce flexibility of syllabus' and school C comments: 'Setting within bands has recently been introduced . . . this has greatly facilitated the teaching of the subject making it easier to gauge the correct pace of the approach with any given group'. Thus it is likely that often mathematics is organised in a particular way for the convenience of the school as a whole rather than through a positive request from the mathematics department.

The secondary mathematics teacher will probably organise lessons in one of three ways: (a) children taught 'traditionally' as a group; (b) children working together in small groups at different tasks; or (c) children working individually at different tasks.

The good teacher, whatever the range of ability of the group facing him, will use a variety of methods. The danger inherent in any method lies in its excessive use. With a group of mixed ability there is a strong temptation to stick exclusively to method (c) with children working from individual work-cards. The result of over-exposure to 'individualised work-card' lessons will be a group of children who find it difficult to communicate mathematics, who are under-stretched and who are sick of the sight of work-cards. (When asked what she did when she had finished a particular card, a pupil replied 'I get another bloody work-card!') Equally the result of excessive use of method (a) will be pupils who show little mathematical initiative and who cannot work on their own. Often lack of confidence or insecurity on the part of the teacher leads to overuse of a teaching style – the good teacher will always be aware of the need for variety.

The question of mixed-ability teaching is discussed more fully in Chapter 5. In his foreword to the Schools Council report on mixed-ability teaching in mathematics (Schools Council 1977), Douglas Quadling wrote:

> Mixed-ability teaching is viewed by some mathematics teachers as a challenge, by others as a threat; but all are agreed that its adoption calls for a reappraisal of classroom organisation, teaching methods and materials.

The challenge and threat are still very much with mathematics teachers – the reappraisal has led to some considerable confusion.

3.5 The Mathematics Teacher

> No externally imposed educational factor can compare in importance with the excellence of a good teacher. He or she must be a motivated, dedicated, born communicator. The system, in as far as it has any hard measuring stick, can make the task marginally easier or marginally more difficult . . . It is the teacher that matters most in the end. (*Guardian*, 1981)

A good teacher of mathematics is, unfortunately, not easily found. This is not the appropriate place to dwell at length on what makes a 'good' teacher, but it is pertinent to consider the qualities we might look for in an ideal secondary school mathematics teacher. He must possess:

(1) a sound knowledge of mathematics to a level beyond that at which he teaches,
(2) a genuine interest in the subject,
(3) an appreciation of the 'grammar' of mathematics,
(4) a full awareness of suitable applications of, elementary mathematics and links that exist with other subjects,
(5) an awareness of a wide variety of resources including books, charts, films, apparatus and materials used in everyday life and within our immediate environment,
(6) a knowledge of at least the broad outline of the history of mathematics,
(7) an understanding of the common problems which children are likely to encounter in learning mathematics,
(8) a liking for mathematical puzzles and access to a good collection of these for classroom use,
(9) an awareness of the particular problems of very slow or very gifted pupils,
(10) a mathematical sense of humour.

Many of these qualities are probably self-evident and incontravertible. Some relate to issues discussed in other chapters – (d) in Chapter 9, (g) in Chapter 4, (h) in Chapter 8 and (i) in Chapter 6.

A mathematics teacher can be forgiven for teaching badly (all teachers have bad days), but he cannot be forgiven for teaching incorrect mathematics; hence items (1) and (3) are a vital part of the repertoire. Some alarming examples of incorrect or ungrammatical mathematics which have been seen in the classrooms of experienced teachers include: 'If set A is bigger than set B, the circle we draw for A is bigger than that for B.' 'If you work out $^{22}/_{7}$ you get the value of π.' 'If $f''(x) = 0$ then $f(x)$ has a point of inflexion.' 'Since $ab = ba$, multiplication is associative.' (Indeed after this statement an argument ensued in which the teacher refused to agree that he misunderstood the difference between associativity and commutativity!) 'Log (-2) + log (-3) = log 6.' 'An angle is a point where two lines meet.'

Although errors of this kind are, it is hoped, few, there are nevertheless many teachers who are sloppy in their presentation of mathematics or who allow bad habits and sloppiness to pass without comment in the work of their pupils. Here are some examples:

Solve the equation: $x + 2 = 7$
Solution: $x + 2 = 7$
$= \quad x = 5$

or Prove that: $a^2 - b^2 = (a - b)(a + b)$
Proof:

$$a^2 - b^2 = a^2 - ba + ab - b^2$$
$$a^2 - b^2 = a^2 - b^2$$
$$0 = 0$$

Alas, such shoddy mathematics all too often is marked correct!

Every secondary mathematics teacher should read at least one good history of mathematics. It is hard to imagine that any teacher would not find much within the history of the subject which could enrich and enhance his teaching. Are not pupils likely to be fascinated by, for example, the Pythagoreans and their number mysticism and the use of figurate, perfect or amicable numbers? Or is there not room in school mathematics for the story (even though it may be legend) of Hippasus, a Pythagorean, who reputedly drowned himself after disclosing the devastating discovery of incommensurables? A mathematics teacher should be ready to deal, from his knowledge of mathematical history, with questions like: 'Why are there 360° in a turn?', 'Who discovered π?' 'Who invented algebra?', 'Who discovered the calculus?' and so on. It is unfortunate that, both in school and university mathematics, little attention is given to the historical aspect of the subject, with the result that most graduates in mathematics have no feel for the way in which mathematics developed and, although they probably know *names* like Euler, Cauchy, Laplace, Leibniz, Newton, Legendre, Pascal . . ., have little idea when these people lived or what they did.

It may appear rather frivolous to suggest that a mathematical sense of humour is an essential prerequisite for a teacher. But if the subject is to be presented in a way which will encourage pupils to think that mathematics can be fun and can be enjoyable, then humour becomes a vital tool for the teacher. Do many teachers know of classics like *Flatland* by A. Square (Abbott, 1974) or Stephen Leacock's essays on A, B and C or boarding house geometry? Yet generations of pupils have sat absorbed and amused as they listened to their teachers reading extracts from works such as these. An awareness of the humour of the subject and a knowledge of suitable sources can be invaluable to a teacher. The following letter, which appeared in the press some years ago illustrates how an alert teacher might seize on a situation which not only provides humour but also provokes some serious mathematical thinking:

> Sir,
> There is an old story of a lorry being driven along a motorway which stopped every few miles to allow the driver to dismount and poke mysteriously about in the back with a long pole. On being asked, not unreasonably, by the police what he was doing, he explained that his vehicle was only a three-tonner whereas he had ten tons of budgerigars aboard. He therefore had to keep them in the air to ease the burden on the axle.
>
> Would the fact that the birds were in the air make any difference to the weight in the lorry? The mathematical geniuses of my acquaintance all seem to differ sharply on this problem.
>
> Yours faithfully,
> John Myers, London.

When secondary school mathematics is taught exclusively by lively, well-informed and appropriately qualified teachers, all the current problems will disappear. Alas, it seems unlikely that the time will ever come.

3.6 Transition to and from Secondary Mathematics

In most areas, pupils transfer from a junior to a secondary school at the age of eleven. In some places transfer may occur at some other age, often 13. But in all instances there is a complete break in a pupil's education with consequent implications for mathematics. Other breaks in continuity occur for some pupils at the age of 16 as they move to a sixth form college and for some when they leave school and move to a university, polytechnic or college to pursue further mathematical studies.

One of the major problems facing a secondary school mathematics teacher is the diversity of mathematical backgrounds of pupils entering his school. In some areas commendable consultation takes place between a secondary school and its feeder schools, but in many cases it is likely that pupils from different junior schools will have studied mathematics consisting of such diversity of content and method that the secondary teacher starts out with an almost impossible task in trying to devise appropriate first year courses. The mathematical experience and background of a junior school teacher is very different from that of one in a secondary school. There is a danger that, in a junior school, pupils will meet advanced (for them) topics in a brief, superficial way – maybe taught incorrectly – so that when they arrive in secondary school they are likely to react with, 'We've done this before', thus destroying the interest and novelty value of a new piece of mathematics. In a group of comprehensive schools, the heads of mathematics, having obtained copies of schemes of work from feeder schools, drew up lists of surprise inclusions and omissions which included the following:

(1) *Surprise inclusions*: Formal treatment of index laws, solution of simultaneous equations; directed numbers; formal establishment of $C = \pi d$, $A = \pi r^2$ and Pythagoras' Theorem; angle properties of a circle; tangent ratio in trigonometry; bearings; arithmetic and geometric progressions; percentage increase and decrease; plotting graphs from equations.
(2) *Surprise omissions*: Use of imperial units.

List (1) was much longer than list (2) and suggests that perhaps junior schools often attempt to be too ambitious with the subsequent creation of problems for the secondary school.

The difficulties facing the secondary school mathematics teacher are illustrated by some of the points made in the following comments of a comprehensive school head of mathematics:

> From one school the information was in the form of a list of topics; the other school stated which textbooks are used in the final year. The first school emphasised that pupils work on an individual basis and so there is a wide variation in the mathematical experience of pupils leaving that school. The other school 'set' their pupils into three groups and the topics covered by the groups vary considerably . . . it is not easy to know which pupils have covered which topics unless the information is given to the secondary school . . . our year one pupils are in mixed-ability groups and we do not have an individualised work scheme – we have neither the financial nor 'time' resources for this. Some pupils will meet very little new work in their first year and some will have already met topics which

> appear as late as our year 4 syllabus. The general feeling of the teachers in my department is that in many topics incoming pupils know 'the rule' but have very little understanding of the concept; they sometimes show impatience when the reason 'why' is being considered.

Although much helpful liaison has been established between junior and secondary schools in many places, there is clearly a problem involved in transition and scope for improved and strengthened links between schools.

The other major problem of transition often occurs as students move from school to university mathematics. Some universities have made great efforts to establish contact with schools but there is still undoubtedly a gap between school and university mathematics and many students find themselves having to struggle hard on arrival at a university. There is a strong case for more contact, and even interchange of staff, between schools and universities. There are, unfortunately many university mathematicians who teach first-year undergraduates who are unaware of what really goes on in school mathematics and who seem unwilling to make the effort to find out. Since the sixth-form mathematics teacher will probably have only a small proportion of any group of students going on to read university mathematics, the onus for improved liaison must lie firmly with the university teacher.

3.7 The Future of Secondary School Mathematics

A secondary school mathematics teacher in the late 1950s, if asked to predict the state of school mathematics in 1980, would almost certainly have not anticipated the sweeping changes of the subsequent twenty years. Equally it is impossible to anticipate the way school mathematics will develop as we move into the next century. As a salutary reminder of the need to be prepared for change, it is worth remembering that a new teacher starting work in 1985 *could* still be teaching in 2025 and that a child starting at an infant school in the same year may well be alive and using mathematics in 2070! In view of the rapidity of change and technological development, one essential priority is clearly that teachers of mathematics should be given ample opportunities to keep themselves up-to-date through appropriate in-service courses and activities. The Cockcroft Report gives strong support to the continuing education and training of mathematics teachers throughout their careers and it is to be hoped that the report's recommendations in this area will not be quietly shelved and forgotten.

The most immediate problems to be tackled within secondary mathematics teaching centre on the content and organisation of mathematics in schools, the use of calculators and computers and the provision of adequately qualified teaching staff.

Content and Organisation

The content of any course should be constantly under review. Alas, the new courses of the last twenty years are already tending to become 'fossilised' in the same way as many of their predecessors. It is vital that school mathematics keeps constantly in touch with the needs of society and employers and at the same time strives to maintain a balance between acquisition of technique, applications and areas of interest. The constraint of external examinations imposes a too firm and unyielding framework for much of the work done in schools and perhaps the responsibility for ensuring constant updating lies both with examination boards (who need to be more flexible and to review syllabuses more frequently) and teachers (who need to show a much greater readiness to devise their own courses).

Sadly many teachers appear to have opted for large scale individual work schemes as an answer to the problem of mixed-ability groups. The result is often a monotonous programme involving boring repetition of type of activity and teaching style. The way ahead in secondary mathematics surely lies in a compromise over ability groupings in schools and a constant *variety* of teaching methods and modes of organisation within the classroom.

Calculators and Computers

No teacher can afford to ignore the implications of new technology. Unrealistically the mathematics teacher is often expected to be the school's computer expert. The emergence of 'Computing' or 'Computer Studies' as a subject in its own right, with its own teachers, must be inevitable. Only then will the mathematician be freed from the added burden of organising and teaching all school computing activities, although clearly any mathematics teacher will want to liaise with computer teachers and make use of their subject in much the same way as at present school mathematics and physics teachers liaise (or at least *should* do).

The calculator has now become an almost universally accepted tool in school mathematics. There is at present a need for teachers to appreciate fully the potential of calculators in the classroom. Doubtless better machines will be available in the future and more uses and applications possible. There is a need for more teacher resources to be produced and for more research to be pursued in this area – some of the work discussed in Chapter 7 provides a good beginning.

Teaching Staff

New methods of teaching, content and technological aids are insignificant when set against the quality of the teacher. It is the individual teacher who will excite, interest, inspire and stimulate or bore, weary and depress his pupils. It is worth spending money on teachers as well as resources and clearly something needs to be done, urgently, to oust unsuitable mathematics teachers from their jobs and replace them from the ever-increasing supply of new teachers and also to set up a far fuller programme of mathematical in-service activity. The present quality of secondary school mathematics teaching is not good; it must improve.

School mathematics has experienced a revolution in the last 20 years. Perhaps it now needs a period in which to reflect and consolidate; such a period is unlikely to be available. The demands on the mathematics teacher will continue to be great and the changes ahead multifarious. A study of the history of mathematical education over the last 100 years suggests that there has never been a period when mathematics teaching was without problems. Today more children than ever before face more mathematics lessons in a climate where changes are taking place with alarming rapidity. Teachers of mathematics need to preserve a calm and careful balance between traditional and progressive ideas and methods and above all need to strive to show children that within mathematics there is something for everyone ranging from fun and amusement at one extreme to immediately direct, practical use at the other.

Bibliography

Abbott, E.A. (1974) *Flatland: A Romance of Many Dimensions*, Basil Blackwell, Oxford.
DES (1980) *Aspects of Secondary Education in England*, HMSO, London.
Freudenthal, H. (1973) *Mathematics as an Educational Task*, D. Reidal, Dordrecht, Holland and Hingham, Mass., p. 64.
Guardian (1981) Leading article 25 June 1981.

Hammersley, J.M. (1968) 'On the enfeeblement of mathematical skills by "Modern Mathematics" and by similar soft intellectual trash in schools and universities', *Institute of Mathematics and its Applications, Bulletin* (October 1968) 66–85.

Hart, K.M. (ed.) (1981) *Children's Understanding of Mathematics: 11–16*, John Murray, London, p. 179.

NCTM (1981) *Priorities in School Mathematics*, National Council of Teachers of Mathematics, USA.

OU (1980) *Mathematics across the Curriculum*, Open University, Course PME 233, Unit 1, p. 22.

Schools Council (1977) *Mixed-ability Teaching in Mathematics*, Evans/Methuen, London and New York, p. 7.

Thwaites, B. (1972) *The School Mathematics Project: the First Ten Years*, Cambridge University Press, Cambridge and New York, p. 213.

Trepte, J. (1981) Extracts from teaching practice notebook (unpublished), University of Durham, School of Education.

4 DIFFICULTIES IN LEARNING MATHEMATICS

Peter Richards

4.1 Introduction

Whether or not mathematics is a difficult subject it is certainly regarded as such by most people. If asked to sum up their view of mathematics at school many people would describe it in terms of one, if not all of the three 'd's' – dull, difficult and disliked. Yet this is in marked contrast to the opinion of a minority that mathematics is interesting, easy and enjoyable. Why this should be and what steps can be taken to put the word 'difficult' in perspective and improve the image and practice of mathematics in schools are questions that occur throughout this book. In this chapter particular attention is paid to what can be learned from studies of children's learning including their attitudes to mathematics, motivation, errors, understanding and rote learning, and the development of mathematical concepts.

There is a great deal of evidence to suggest that school mathematics has been made unnecessarily difficult and that the teaching of the subject has run contrary to many general principles of learning and mathematical learning theories which acknowledge the developing structure and abstract nature of the subject. At secondary level, in particular, failure has often been more obvious than success, rote learning more apparent than thinking and problem solving, and knowledge more important than personal exploration and self-development. At primary level the movement towards child-centred methods of learning brought a refreshing change in approach but too often psychological theories were over-enthusiastically compressed into simple principles of limited application, leaving teachers wondering to what extent they should structure and consolidate work and whether it was possible to accelerate concept development by verbal examples and discussion.

There have been some signs of late of an over-reaction in favour of a 'back-to-basics' approach, but the evidence of the recent HMI, APU and Cockcroft reports suggests that mathematical competence should extend well beyond the ability to perform basic techniques in narrow situations. If some difficulties in learning mathematics have arisen from an excess of active open-ended work and a lack of mechanical skills, more serious difficulties are likely to emerge if the process is reversed. Passive methods of learning are not only limiting mathematically but they weaken pupils' self-confidence and the essential process of thinking for themselves.

The mathematics teacher's responsibility for his pupil's personal as well as mathematical development is not acknowledged as much as it should be, and if not always stated explicitly it is to be regarded as the underlying philosophy of this chapter.

4.2 Attitudes and Emotional Reactions to Mathematics

Considering the amount of work devoted to studying and writing about children's intellectual development in mathematics it is surprising to find so few studies dealing with the presence of emotional and attitudinal factors. The almost total preoccupation of Piaget and his co-workers with the cognitive aspects of children's development has been one reason for the balance of attention falling as it has done, and though it is tacitly assumed that a person's emotional reaction to mathematics can be a crucial factor in their learning, this factor has not been emphasised. Piaget's own references to the domain of emotions acknowledge the duality of the cognitive and the affective and he clearly believes that many of his cognitive findings are mirrored in the affective area. 'Affective life', he notes, 'like intellectual life, is a continual adaptation, and the two are not only parallel but interdependent, since feelings express the interest and the value given to actions of which intelligence provides the structure.' (Piaget, 1973).

There are therefore two aspects to the role of feelings to be noted, first that they are an important and integral part of learning and secondly that children's feelings, like their thinking processes, pass through various stages of development. Emotionally, as intellectually, children develop from an ego-centric stage where basic feelings such as pleasure, success, fear or failure are predominant, to a formal operational stage associated with questions of values and judgements.

Emotional development will therefore have a bearing on the type of satisfaction a pupil can derive from his mathematical experiences. Young children, for example, can derive a great deal of satisfaction from getting right answers, but when the intellectual challenge of the work is limited we need to recognise that the satisfaction comes more from the teacher's approval than from any intrinsic motivation in the exercise. At an early age it is difficult to separate pupils' emotional reaction to a task from that derived from the teacher, but as they get older the balance changes and they will be less prone to persevere at a task simply to 'please teacher'. The task has to be made more interesting for its own sake and the type of teacher approval also needs to become more 'mature' and incorporate responses that obviously value pupils' opinions and respect their views.

One of the reasons why mathematics arouses such strong feelings may well be that the quality of personal communication between teacher and pupil is neglected in favour of what we may be tempted to see as the value-free nature of mathematics. We may justify categorising answers as right or wrong in the belief that they are simply an objective indication of arithmetic skill alone, but for pupils they are often associated with strong feelings of warmth or disapproval respectively. As Anthony (1976) points out human beings are far more important sources of satisfaction and distress than inanimate objects and are liable to have strong positive or negative feelings attached to them. Our own attitudes to a young child's mathematical performance may therefore have more influence on his future development and emotional reaction to the subject than the special characteristics of the subject itself.

By the time the pupil has reached an age when he can make a rational assessment of his feelings and accept and analyse failure and critical comment in a non-personal way, the scales are often heavily weighted against the subject, and attitudes once formed are notoriously difficult to change. An image of oneself as being poor at mathematics could well spread into a more general assessment of oneself as a failure but, though it does not help the image of mathematics, we can in some ways be glad that there are safety mechanisms to prevent this. As they get older pupils and adults who have 'failed' at mathematics learn to preserve their self-image by devaluing the importance of the subject or even by taking positive satisfaction in belonging to groups with 'anti-mathematics' opinions. One unfortunate consequence is that these values can be transmitted to others, and the teacher often has an upward struggle to improve the difficult and disliked image of mathematics that some children acquire from their parents and bring with them on their first day at school.

Although emotional development may well follow the same broad stages that Piaget has demonstrated for thought processes it is likely to be less ordered and subject to greater variations. It is a less conscious activity than intellectual development, and although a child should be continually adapting his emotional responses to his environment by processes of 'assimilation' and 'accommodation' similar to those he uses in the formation of his intellectual concepts, there is a good deal of evidence in psychological literature to show that children and adults can remain 'fixated' in their emotional responses at a level which does not develop with experience.

When a child meets new experiences of an intellectual kind he is likely to have to face up, consciously, to certain incidents which conflict with his existing concepts. In doing so he learns to modify or change his ideas to come to terms with the real world. Emotional concepts are not so consciously challenged however and immature responses can be more easily maintained. It is plausible, therefore, to suppose that if felt deeply enough, unpleasant early experiences of mathematics may result in serious emotional difficulties with the subject which may persist into later life.

The depth of feeling that successful adults can still have about their difficulties in school mathematics comes over very clearly in a recent book by Buxton (1981). Notwithstanding the selected nature of the group studied, much of its evidence is disturbing and thought provoking as the negative effects of school experiences are convincingly portrayed. Buxton claims that 'panic' is not too strong a word to describe the reactions that people can have when confronted with having to 'do' some mathematics, often remembering pressures of time and the threatening displeasure of an authoritative teacher, or the embarrassment of having to demonstrate ones inadequacy to the rest of the class. Lack of understanding in mathematics can be so easily and so devastatingly demonstrated that children frequently experience feelings of failure and discomfort and react accordingly.

Most of the people selected by Buxton could remember reactions like, 'Oh, my God, I'm going to make a fool of myself', and described how they associated mathematics with fear and trembling, or a complete detachment. Physical symptoms of panic and despair were also reported, cold sweats, clammy palms and 'a lump in the throat – feeling that you could get some release if you could but cry'.

The memories were not confined to very early experiences such as the demand to remember 'seven times seven'. One comment revealed that understanding was switched off and panic switched on when it came to quadratic equations, and another when the first 'θ' appeared. Nor were the responses conditioned only by experiences at school. Failure to please parents by instant and correct answers to questions on tables, and the feelings of parental disappointment that were obvious as a result, were recognised as having serious effects.

Pressures on some of Buxton's adults to give instant and correct answers to mental arithmetic calculations still provoked genuine anxiety and growing panic. One person beginning to react in this way recovered only when encouraged to take her time and when the 'teacher' conveyed clearly that there were going to be no judgements or recriminations. The reaction to this removal of tension was a spontaneous regression to a feeling of 'Thank you, thank you for not being cross.'

It is clear that Buxton's book is not meant to illustrate that mathematics teachers are being sadistic, but rather that, however kindly they are, teachers often do not realise that mathematics learning can arouse such a depth of feeling. it was interesting to note that, in spite of their difficulties, some of the adults interviewed by Buxton had nevertheless quite liked their mathematics teachers and had recognised that the teacher really had wanted to help them. If after being helped one still fails to succeed however, this can become another source of discomfort and guilt. It is difficult for teachers to keep from a child the disappointment they may feel at his lack of progress, but if they do not do so then their help in future may be even less effective. The child may well begin to build barriers to future communication for fear that if he remains unsuccessful after further help he will disappoint the teacher even further. Avoidance of the painful is a natural reaction. There are unfortunately plenty of signs of this in the mathematics classroom. Pupils begin to look bored, their eyes glaze over at attempts to explain, they look elsewhere, they begin to fiddle with pens and pencils and generally communicate studied disinterest. How do we react to such signs? Hopefully with patience and understanding, but often, understandably, by showing annoyance and by putting further pressure on the child to attempt what he is finding impossible in his present state. It is not easy to get through this barrier and it will not be achieved overnight. The initial reasons for pupils' lack of understanding may well be cognitive ones, but it is important to recognise the emotional barriers that can be set up as a result – barriers that can be erected to cut off the pupil from what he sees as the 'threat' of mathematics.

Recognising that children's difficulties with mathematics may be largely emotional, and being sensitive to how they might arise, can help teachers prevent a good deal of misery and put them in a position to help those who may already have emotional problems with the subject. Talking to pupils about their reactions and getting them to articulate their feelings is a crucial step in remedial work.

4.3 Motivation

Mathematics learning is likely to be most effective if the pupil's experiences are challenging enough to arouse his interest and active participation without being so threatening as to cause an avoidance reaction. Some degree of motivation is essential if learning rather than conditioning is to result and the conceptual difficulty of much mathematics learning demands a correspondingly high level of 'active thinking' if real development is to take place. Teachers use a wide range of methods to encourage children to work, but it would be wrong if these relied solely on extrinsic rewards or punishments, or on the authority of the teacher rather than on the nature of the task and its relevance to the pupil. If mathematics is important we ought to be able to convey this and provide material which pupils can find personally rewarding and satisfying. One of the keys to human motivation is that the mind is naturally curious, or as White (1959) suggests is characterised by a drive of an intellectual nature – 'competence motivation' – which stimulates active exploration and experiment. As Piaget argues the latter is essential in concept development and it is also the way in which children gain their personal identity and self-realisation.

Many of the problems of motivating pupils arise from considerations well beyond the classroom, such as the value placed on schooling by parents and peers, but there is a great deal that can be done by the teacher to capitalise on the positive features of his subject and the characteristics of intrinsic motivation which Bruner (1966) has termed the 'will to learn'. Emphasising the value to learning of intrinsic motivation and echoing some of the characteristics outlined by White, Bruner suggests that it is so natural as to be involuntary and is characterised by curiosity, a search for competence, modelling on identifiable people, and the need to respond to and co-operate with others. Each of these motivational factors can be encouraged in the mathematics classroom.

Curiosity can be stimulated by mathematical materials, games, puzzles, patterns and problems, and there is room for exploration and experiment. The level and type of stimulation needed to encourage pupils' curiosity obviously depends on their ability and past experience, and the guiding principle is that it is the degree of incongruity between previous knowledge and the new situation which is intrinsically motivating. The incongruities have to be recognised as such and not be so so great as to appear totally unrelated to previous experience. They need to be stimulating without being overwhelming.

The achievement of 'competence' is one aspect that is so common in the mathematics classroom that it can be overdone unless the exercise is suitably challenging. Practice followed by success is a rewarding experience, but the activity needs to be intellectually meaningful. If the balance of challenge and reinforcement is wrong even the achievement of correct results can become boring. Even rats when offered a short, direct path to food, and a longer more variable and less direct pathway involving a 'search' for food, have been shown to prefer frequently the more difficult, but more 'interesting' route! (Hebb, 1955).

The process of 'modelling' is another motive for learning as pupils seek identification with a person or a set of values to which they aspire. It underlies the importance of the teacher's own interest in and curiosity about his subject as a model for his pupils. He can also communicate the essential relevance of his subject both as a form of knowledge and in terms of its practical relevance to employment and everyday life. Mathematics is a particularly potent subject for exploitation in both respects.

The principle of 'reciprocity' is a basic social motive to respond to and co-operate with others. It does not imply conformity but the self-reinforcement of working with others and sharing common goals. Project work in mathematics and real problem solving exercises where the pupils work as a class or in small groups are common at primary level, but the potential of mathematics for this form of enquiry is often neglected at secondary level.

4.4 The Development of Mathematical Concepts

Background

The results of some detailed investigations into pupils' mathematical achievement have revealed that pupils throughout the secondary school and in further education have considerable difficulty with certain types of mathematical skills and concepts. The work of Rees at Brunel University (e.g. 1973), the Concepts in Secondary Mathematics and Science (CSMS) team at Chelsea College (e.g. Hart, 1981; Brown, 1979) and the APU surveys show clearly that mathematics, judged by a child's performance on test items, is a difficult subject for the majority of pupils.

Although, as one APU survey (1980) observed, pupils could

usually carry out many basic mathematical operations, their facility decreased sharply when the concepts were probed more deeply or when their knowledge had to be applied in more complex or unfamiliar contexts. As one tester pointed out pupils often 'did not recognise the problem and the algorithms as the same'. The futility of encouraging pupils to rely on algorithmic learning without sufficient understanding of the concepts involved is also demonstrated by the CSMS results. Out of a sample of over 1,000 children aged eleven to twelve, for example, less than one third of the children could think of any practical problem they could solve by multiplying 56 by 28, and between 30 and 40 per cent of twelve-year olds identified expressions such as $391 \div 23$ and $23 \div 391$ as the same.

The CSMS investigations were conducted within a broadly Piagetian framework of development and though they were able to demonstrate only an approximate hierarchy of concepts across the whole curriculum, with some pupils performing at different levels on different topics, they do suggest that there are broad levels of understanding; and that concept formation in secondary schools is still largely linked to concrete representations. Formal reasoning is reached in only a small proportion of cases.

The answer to Rees' question whether teachers underestimate the difficulty their students have with concepts and skills is almost certainly that they do; and the conclusion of most recent surveys of children's difficulties has been that 'It certainly seems likely that children's problems with learning mathematics are rooted in their difficulties in forming conceptual structures' (Brown, 1979).

Concept Formation

Learning in mathematics depends very much on the pupil's grasp of earlier basic concepts which should not be inferred from the ability to perform routine tasks in a narrow and limited context. A pupil may learn, for example, to divide, cancel or subtract in particular contexts, but this carries no guarantee that he can appreciate the meaning of the exercise in the way the teacher might expect.

One of the main results of Piaget's work has been his clear demonstration that the developing child thinks in a very different way from an adult and that we should not assume that because a pupil's speech or actions mirror our own that he is able to think in the same way as we do. In his experiments Piaget deliberately attempted to devise situations which would delve into the real beliefs that exist in a child's mind and his theories of intellectual development have had a major influence on the development of school curricula and classroom organisation in the primary school. In view of the evidence of difficulties with concept formation well beyond the primary stage more note of his work might be taken by teachers at secondary level. During his life, 1896–1980, Piaget did more than anyone to illuminate the ways in which a child develops his faculties for thinking, reasoning and understanding, and if we are to overcome many of the difficulties in learning mathematics we need to understand something about how these capacities grow in the child. Only the barest details can be given here but there are many books summarising his theories (e.g. Flavell, 1963) and many more reporting on research which has arisen from his work (e.g. Collis, 1975). There is no doubt that some of his theories were too rigid and some of his interviews misinterpreted, in spite of his attempts to use the 'art of questioning' to capture the real nature of a child's ideas, but these do not detract from the overall value to teachers of a study of his work.

Whatever conclusions are finally made about the details of Piaget's theories he has established not only that children's thought processes are very different from those of adults, but has provided a structure which indicates both how and in what sequence concepts are developed.

The development of children's concepts, he maintains, is dependent on mental activity that takes place as the child experiences and interacts with his environment. Experience is the key word, and to develop their concepts effectively he insists that children need to be given opportunities to perform their own physical actions. Until they are able to reason abstractly therefore he argues that children must have real and relevant practical experiences if they are to build up and 'internalise' a concept. Verbal representations play only a very secondary role in his theory, though he does note the value of social interaction and its associated exchange of viewpoints and discussion in the period of formal operations.

Verbal demonstrations to pupils in order to develop their concepts is therefore considered a futile exercise unless it is related to situations in which the child himself experiments. Concept development can be directed however by providing pupils with materials that exemplify the relationship which, when abstracted, becomes the concept. Children may, for example, abstract a concept of 'triangle' from experience with different shapes, and use the same property to recognise triangles of different size, colour or shape – until they need to modify this concept at some stage to consider triangular prisms. To make concepts fully operational the teacher should present pupils with as great a variety of situations as possible which exemplify the concept. It is surprising how many pupils lose their concept of trigonometric ratios in a right-angled triangle if the triangle is stood on end!

The second aspect of Piaget's theory that has an influence on the way we teach mathematics is his description of four main stages of intellectual development through which children pass on their way to forming adult cognitive structures. The main value of his stage theory being the characteristic ways of thinking that are associated with each stage. Piaget (1962) describes the stages thus: 'I shall distinguish four great stages in the development of intelligence: first, the *sensori-motor* period before the appearance of language; second, the period from about two to seven years of age, the *pre-operational* period which precedes real operations; third, the period from seven to twelve years of age, a period of *concrete operations* (which refers to concrete objects); and finally after twelve years of age, the period of *formal operations*, or propositional operations.'

The model is useful to teachers as a broad guide to the way in which pupils develop and as such has implications for the way in which mathematics should be geared so that the complexity of the subject matter is matched to the conceptual abilities of the child. This is more complicated than it sounds however, for there is considerable variation between individuals of the same age group, and even more important between the conceptual structures of an individual in different contexts. With one type of material for example a child may be able to demonstrate an understanding of conservation or reversibility but not in another. It would be impossible therefore to apply a blanket label to his stage of development which would automatically dictate certain suitable learning experiences. While it is important to recognise the qualitatively different stages that Piaget has shown to take place in concept development they can only provide very flexible guidelines for curriculum planning.

Bishop (1976) in a review of Krutetskii's work suggests that in the long run it may be more profitable to consider the development of concepts, not in general stages, but in terms of the abilities which Krutetskii (1976) maintains can be seen first in embryo in primary grades and then again as they develop gradually as the pupil gets older. Krutetskii puts more emphasis than Piaget on the effectiveness of instruction, but acknowledges that the process of development can only be compressed within certain limits.

Piaget does not however suggest that his stages are delineated by definite jumps in ability, and it would be wrong, for example, to assume that children move automatically from a concrete to a formal operational stage of thinking when they reach the age of twelve. The fact that this age is attained during the first year of secondary education in this country, however, has implied some support for a teaching approach at secondary level which is away from the concrete and towards the verbal. For many children this will be out of step with their level of cognitive development, and in an attempt to keep up with the work they will be forced to rely on techniques which will give them only temporary superficial success. It should also be noted however that, in the other direction, there may be pupils at primary schools who are capable of formal reasoning with verbal propositions and they should not be confined to purely descriptive work in the mistaken belief that Piaget's theory implies that they could not possibly have reached the formal level.

The ages given by Piaget are merely guidelines but to be useful they need to point to the stage of development of the average child. Lovell (1978) suggests they are more indicative of the development of able pupils and that for pupils of average ability, formal operational thought begins around 13 to 14 and does not come to maturity until around 17 or 18 years of age. In a CSMS survey quoted by Brown (1979) only about 10 to 15 per cent of children at the age of 12 gave responses classified as reaching the early formal stage, and of these only a small proportion reached the late formal stage. Even at 16 only about 25 to 30 per cent were classified at the early formal level or above. It appears, as Brown concludes, that the vast majority of secondary school children are progressing through the concrete operational rather than the formal operational stage. This view is also borne out by Collis (1975) who sees the concrete operational level extending from 10 to 15 years of age with formal operations arriving only at 16 plus. He designates the 13 to 15 plus stage as one of 'concrete generalisations', for although the pupil is still tied to reality Collis suggests that he is seeking generalised solutions within this restriction. More attention to such a stage may be very useful in planning work at this level.

Although Piaget excludes verbal explanations alone as a means of helping young children to form their concepts he acknowledges the increasing value of language and discussion as the formal operational stage is approached. Skemp (1971) suggests how language can be used to convey 'experience' and maintains that provided explanations rely only on concepts that the person has already abstracted it is possible to convey 'lower-order' concepts by verbal definition. In the teaching of mathematics however he notes that the movement is almost always

towards higher and more abstract concepts and that verbal definition is not sufficient for this purpose. What is possible however is that the teacher can provide verbally a collection of suitable experiences and examples to help promote the development of the concept in the same way as Piaget has pointed to the role of practical experience for younger children. Once again the examples used must involve only those concepts which the learner can already understand. This is the crux of the problem for many learners as all too frequently they do not understand the work on which new explanations are based.
It means that if verbal methods are used teachers will have to be very aware of the need to structure their explanations strictly in accord with their pupils' levels of understanding if the children are to be able to progress mathematically.

Curriculum Implications

For the teacher of mathematics the problem is to provide mathematical experiences for his pupils to suit the state of development of their existing concepts and to fit his method of presentation to cater for the pupils concrete or formal level of thinking. At the same time he has a responsibility to foster the pupil's ability to analyse new material for himself so that he can develop his own concepts, independently of the teacher, in ways most meaningful for him.

Piaget emphasises the latter aspect since his theory revolves about the pupil's active structure of his own experience, but our school system is geared to a greater degree of direction by the teacher than this implies and few teachers would have the resources, courage or freedom to take this principle literally. Instead Bruner's idea of a spiral curriculum with material structured and later restructured to provide experiences at the right level of pupil development has been appealing, though the individualised learning procedures that this implies have not always been successful in maintaining the zest for learning, opportunity for discovery, and pupil autonomy that Bruner also emphasises (Bruner, 1966).

A good deal can be learned from curriculum developments at primary level which have adhered to the principles of structured learning whilst maintaining pupils' personal activity and exploration. Indeed Armstrong (1975), basing his views on those of Hawkins (1973) argues that the emerging primary school tradition with its emphasis on the personal experience of the individual child should be built into the rationale for a similar education in secondary schools. Although acquiring knowledge by 'primary methods' is regarded by many teachers as less 'intellectual' than the subject discipline approach of the secondary school, Hawkins emphasises that the ability to reorganise subject matter and weave together real experiences to 'resonate with the thought processes of those he teaches' is a major intellectual and practical achievement and demands a 'wide, fluent and reflective grasp of his subject matter'.

It will be some time before the secondary school curriculum is reorganised to cater for the type of involvement of teachers and pupils in the real problems advocated by Hawkins, but it is one way in which we might provide motivation and opportunity for individual development, particularly for pupils who find mathematics difficult. For such pupils more will be achieved by an approach which enables them to understand what they are doing at a realistic and relevant level than by an 'academic' method which is likely to result in a failure to grasp higher mathematical concepts or, at best, to rely on a rote learning of routine text-book skills without the ability to apply them in practice.

To encourage children to work at real problems is one of the aims of a recent course for teachers produced by the Open University entitled *Mathematics Across the Curriculum* (OU, 1980). It emphasises the importance of developing children's mathematical skills and ideas in exploring real situations which are personally relevant to them, and reveals more opportunities for exploring real problems in the classroom than one might think if conditioned to see mathematics solely in terms of syllabus content to be taught in isolation from real events. Real applications of mathematics have been the subject of the Schools Council's Applicable Mathematics Project, and many useful ideas for mathematical projects are given in a booklet *Pupils' Projects* produced by the Mathematical Association (1980). There is a difference however between projects which are real in terms of the learner's own immediate environmental experience and those which demonstrate the wider applicability of mathematics.

Dienes (1973) has suggested a framework for the mathematics curriculum which emphasises Piaget's principle of active experience and covers six stages of learning, beginning with exploration and ending with a set of mathematical structures. An initial 'environment' is specially designed to embrace certain features from which the required mathematical concepts can be gradually abstracted and organised. The first stage, Dienes maintains, is of outstanding importance as it incorporates the basic principle that learning is a process by which a person actively explores and adapts to his environment.

The fact that he gives his logic blocks as a possible environment and describes stage one as a 'play stage', may give the impression that his method is only applicable to young children, but the 'environment' could consist of a set of Skemp's verbally-constructed exemplars of a concept and in this way the theory could be made applicable to pupils of all ages. The construction of suitable 'environments' for all the mathematics that pupils, especially able pupils, may need to learn would be a daunting task however, though it may be an effective means of developing the concepts and abstract reasoning abilities that such pupils will need in order to gain access to the vast store of accumulated mathematical knowledge.

Many learning difficulties, Dienes maintains, are the result of teaching methods which proceed in the opposite direction, and the structure of many secondary school text-books bears out his contention. The interesting examples usually come at the end – after one has attempted to develop a formal mathematical system to answer the problem. The better examples would be a useful means of creating the environment from which the mathematical structures might be built up.

Evaluation of certain aspects of a secondary (11 to 13) mathematics course designed in accord with Piagetian philosophy (The South Nottinghamshire Project) is being carried out at Nottingham (Bell, 1980). It is designed to examine the effectiveness of a set of teaching methods derived from Piagetian theory and also the pupil's attainment of general strategies of mathematical activity. As most investigations of Piaget's theories in action have taken place at primary level the results of this work may clarify the problems of concept development in those areas of the secondary curriculum which have been highlighted as proving difficult by the researches at Brunel and Chelsea. The Nottingham study is making use of criterion tests designed by the CSMS team at Chelsea, and it is worth noting that these are now published by the National Foundation for Educational Research.

4.5 Aiming for Understanding?

Skemp (1976) has suggested two forms of understanding which help to clarify some of our aims in teaching mathematics. On the one hand we want children to demonstrate that they can perform a mathematical skill and at the same time we usually want them to understand what they are doing. These aims are distinguished by Skemp by adopting the labels 'instrumental understanding' and 'relational understanding'. He notes that until recently he would not have regarded the former as understanding at all but acknowledges that, though characterised as 'rules without reasons', the possession of such a rule, and the ability to use it, is what many pupils and teachers mean by 'understanding'. 'Relational understanding' on the other hand is described by Skemp as 'what I have always meant by understanding . . . knowing both what to do and why'.

Amongst advantages which might be claimed for instrumental and relational understanding respectively Skemp includes the following:

(1) *Instrumental Understanding*
(a) It is usually easier to understand within its own context. (Consider for example the difficulty of understanding 'relationally' topics such as multiplication of two negative numbers or dividing by a fraction.)
(b) The rewards are more immediate. (It's nice to get a page of right answers.)
(c) It can give a correct answer quickly and reliably.

(2) *Relational Understanding*
(a) It is more adaptable to new tasks.
(b) It is easier to remember. (Many otherwise separate rules can be seen as a connected whole.)
(c) It can act as a goal in itself.
(d) It encourages the learner to intellectual exploration. (By seeking relational understanding in other areas and on their own initiative.)

Skemp goes so far as to suggest that there are two effectively different subjects being taught under the same name 'mathematics'. If this is so two questions arise. First, 'Does this matter?' and secondly, 'Is one kind "better" than the other?'

It is unlikely that neat answers will emerge to the second of these questions. There is still wide discussion about the validity of Skemp's categories and what we mean by understanding, and there is the wider question of how the immediate goals of mathematics teaching are related to the whole purpose of education. The view taken in this chapter is that Skemp's categories are useful in clarifying teaching goals and that an emphasis on relational mathematics would avoid many of the learning difficulties that have arisen from a reliance on instrumental mathematics.

With regard to the first question, whatever ones views about which type of approach to mathematics learning is preferable, one has to recognise the difficulties that can be caused by a mis-match between the goals of pupil and teacher. As Skemp emphasises the mis-match can occur in both ways: 'Pupils whose goal is to understand instrumentally taught by a teacher who wants them to understand relationally', and 'the other way about'. It can also be a mark of contrast between the aims of different teachers even within the same department, and perhaps more commonly, between the teacher's expressed aims and the method he subsequently adopts when faced with the reality of the classroom.

A further source of difficulty identified by Skemp is the mis-match that can occur between the relational aims implicit in many modern mathematics texts and the instrumental teaching approach which is often maintained by teachers used to more traditional syllabuses. Ideas such as sets, mappings and functions have been introduced to help foster relational understanding but if taught instrumentally, Skemp suggests, they will probably do more harm than good.

There needs to be some correspondence between content and methods and if constrained, by narrow previous experience or school organisational variables, to instrumental teaching, teachers need to chose their material accordingly. It is interesting to note that Skemp interprets teachers' difficulties in adapting their teaching methods to suit the new syllabuses as a predictable difficulty in accommodating their existing conceptual structures without adequate experience.

4.6 Discussion and Pupils

In addition to the information that has been collected about which skills and concepts in the curriculum cause difficulty, the efforts at diagnosis have led to a healthy emphasis on individual discussion with pupils which is reminiscent of Piaget's 'clinical method' of questioning. It is also the approach used by Krutetskii (1976) in his investigation of mathematical abilities and, as he maintains, leads to a knowledge of the process used in the performance of mathematics which is for many purposes more useful than the result itself. The high level of difficulty which has been shown to exist suggests that time spent with pupils discussing their methods and analysing the reasons for their mistakes will be very profitable, not only for the pupil but for the insight it would give the teacher into pupils' thought processes.

Magne (1978) for example notes that discussion with pupils experiencing difficulties suggests that even with less able pupils errors are made because thinking is taking place rather than because of an absence of thought. Pupils attempt to apply rules and ideas learnt in different contexts without fully understanding their limitations and lack of generality. They may for example 'take the square root' in cases such as $x^2 + y^2 = z^2$, with predictable results! Or they may see some sense in their answers to both the following:

$$\begin{array}{r} 63 \\ +\,28 \\ \hline 91 \\ \hline \end{array} \qquad\qquad \begin{array}{r} 63 \\ \times\,28 \\ \hline 144 \\ \hline \end{array}$$

For children with these sorts of difficulties 'more practice' without proper remedial work is likely to strengthen rather than reduce the error pattern and result in even greater difficulty. It is worth noting that while more able pupils may be able to modify their methods as a result of class demonstrations, the less able will almost certainly need individual attention.

Clements (1980) reports on a constructive method for helping teachers analyse childrens' errors in verbally-presented problems and the approach he suggests can be useful in encouraging discussion of pupils' difficulties. Dealing with verbal problems that led to one arithmetical calculation (one-step problems) he described data on pupils' errors in terms of five categories; reading the problem, comprehending what to do, transforming from words to a mathematical procedure, applying the necessary skills and finally delivering the answer. Failure at any level would lead to an incorrect result (other than by chance). The results confirmed the suspicion of many teachers that a major source of error in verbally-presented problems is due to reading, comprehension and transformation difficulties, i.e. they take place before the stage where any mathematical skills are needed. With low-ability pupils (aged ten to eleven) 35 per cent of the errors were reading errors alone, and with a complete sample of twelve-year olds, 38 per cent of their errors were at or before the transformation stage. With problems needing more than one step the percentage increased to 45.

In addition between 21 and 35 per cent of errors appeared to be outside the hierarchial scheme and were accounted for in terms of carelessness or lack of motivation. Being realistic another category was provided to acknowledge that teachers can set ambiguous questions, or pupils can give justifiable answers other than the one intended. It is refreshing to find Percy's answer 'one' being recognised as technically correct in response to the question, 'I have 24 lollies and I want each child in the picture (twelve children) to have the same number of lollies. How many lollies will I give to each child?' Asked how he obtained his answer Percy explained, 'I would give each child one lolly and keep twelve for myself'!

Only a limited amount of information about children's errors can be obtained from their written answers, and the simple 'interview' procedure reported by Clements is a useful basis for discussing and elucidating their difficulties. The teachers would ask a pupil who had given an incorrect answer to try the question again and to respond to the following questions or requests:

(1) Please read the question to me. If you don't know a word leave it out.
(2) What is the question asking you to do?
(3) How are you going to do it?
(4) Show me what to do to get the answer. Tell me what you are doing as you go along.
(5) Now write down the answer.

In their attempts to see as many pupils as possible during a lesson teachers frequently succumb to the temptation to 'overcome a pupil's difficulties' by picking up a pen and doing the work for him. The procedure outlined above may encourage more discussion with individual pupils, and it can also be used in classroom discussion. Error analyses do not themselves prevent further difficulties but they provide the essential information on which the teacher can then proceed.

It is also important to realise that not only is discussion in mathematics lessons a means of correcting errors and misconceptions, it is also a fundamental factor in generating a constructive and creative approach to mathematics learning. Brissenden (1979) suggests that it is the most enjoyable and most important part of teaching mathematics, and outlines ways of managing discussion as it proceeds through questions, hypotheses, methods of investigation, modification and evaluation. Discussion or simply 'talk' with pupils is also the means by which the whole relationship between teacher and pupil is built up.

4.7 Learning Disabilities

In spite of their own efforts and all the teacher can do to help them, there are some children who fail to progress or constantly underachieve in mathematics whilst reaching a high level in other subjects. With such children normal teaching procedures and remedial strategies even of the most enlightened type do not seem enough.

A number of children who fall into this category are taking part in a study at Bath as part of a project investigating the incidence of specific mathematical underachievement and a disorder of mathematical functioning known as *dyscalculia*. (May, 1980; Blane, 1980; Richards, 1980). Pupils are nominated by their teachers if they are underachieving specifically in mathematics for no accountable reason. They vary from an eleven year old who scored in the top 6 per cent on an ability test but consistently scored almost nothing on a mathematics test, when her predicted score (on the basis of her general ability) was 77 per cent, to a 17-year old who very successfully passed all her O-level subjects except mathematics and who, after three further attempts and considerable help has not progressed beyond grade 3 CSE. Similar discrepancies between high general ability and low mathematical performance are referred to by Johnson and Myklebust (1967) who describe dyscalculic pupils with a difference between verbal and non-verbal intelligence quotients as wide as 72 points!

Such pupils may, for example, be able to demonstrate an understanding of the structure of numbers and number operations and yet still have a total lack of ability to remember number bonds or multiplication tables; they may know the names of and recognise geometrical shapes and yet be unable to draw them; or they may explain how the number system is based on place value but be unable to arrange the digits of a number consistently in their right places. In order to cope with these problems teachers are likely to need the co-operation of specialists, though the first step is to be aware that such difficulties can exist. It is surprising how pupils can conceal their difficulties and get by with alternative strategies – such as counting on their fingers or copying from their neighbours! It is also unfortunate that we sometimes accept the common 'I just can't do mathematics' excuse without looking more deeply into the reasons why.

As suggested earlier in this chapter poor teaching, lack of motivation, inadequate learning of basic concepts, or serious anxiety are likely to be the cause of most mathematical underachievement, and one needs to look first of all at these areas before jumping to the conclusion that an underachieving pupil has some neurological limitations to achieving his mathematical potential.

It is possible, however, that a child may be prevented from performing mathematical operations, particularly in number work, by a neurological disorder, such as dyscalculia, which Kosc (1974) describes as 'disorders of mathematical ability which are a consequence of heredity or congenital impairment of the growth dynamics of the brain centres'. Unfortunately there are so many inconsistencies amongst definitions, methods of identification and suggestions for remedial work that at present there does not appear to be a well enough defined method of diagnosis to provide the classroom teacher with a useful and effective tool in identifying such disorders. The major difficulty is that the term dyscalculia is often being used to describe the *result* rather than the *cause* of mathematical failure and this is not only unproductive but potentially harmful. It is one of the problems that has led to a good deal of the controversy over labelling certain children as dyslexic and teachers of mathematics should be alert to the same danger with the concept of dyscalculia. It would therefore be unwise to put all specific mathematical learning disabilities under the title of dyscalculia, yet we need to make allowance for the fact that some difficulties in learning mathematics are neurologically based and that amongst pupils who show serious underfunctioning in mathematics while progressing well in other subjects there may be some who should be diagnosed as belonging to this category. If a specific enough condition can be identified proper remedial work can be planned to cope with it.

In one study of children in this category, Weinstein (1978) concluded that dyscalculia is identifiable in terms of neurological disability but suggested that normal remedial procedures should still be effective in coping with the disorder if based on a developmental lag of about four years. It would be convenient if this were the case, but it seems an optimistic assessment of the problem, and the study in question did not test this hypothesis. We need a good deal more information before conclusions can be made about the concept, but it seems more likely that some causes of childrens' mathematical underachievement will be identified that are beyond the reach of normal remedial procedures.

For the vast majority of children however, this chapter has suggested that there is a great deal of learning theory and information about concept development to enable teachers to cater for their pupils in a way which should avoid or overcome many of their potential difficulties.

Bibliography

Anthony, E.J. (1976) 'Emotions and Intelligence', in V.P. Varma and P. Williams (eds.), *Piaget, Psychology and Education*, Hodder and Stoughton, London.
Armstrong, M. (1975) 'Comprehensive Education and the Reconstruction of Knowledge', *Forum*, vol. 17, no. 2, 40–4.
Assessment of Performance Unit (1980) *Mathematical Development: Primary Survey Report No. 1*, HMSO, London.
Bell, A.W. (1980) 'Research on Teaching Methods in Secondary Mathematics'. Occasional Paper. Shell Centre for Mathematical Education. University of Nottingham.
Bishop, A.J. (1976) 'Research: Krutetskii on Mathematical Ability', *Mathematics Teaching*, *77*, 31–4.
Blane, D. (1980) 'Specific Mathematical Failure in Children, its Identification and Diagnosis', *PMEW paper*, Chelsea College, University of London.
Brissenden, T.H.F. (1979) 'Teacher-pupil Discussion in Mathematics', *Mathematics in School*, vol. 8, no. 3, 29–31.
Brown, M. (1979) 'Cognitive Development and the Learning of Mathematics' in A. Floyd (ed.), *Cognitive Development in the School Years*, Croom Helm, London.
Bruner, J.S. (1966) *Toward a theory of Instruction*, Harvard University Press, Cambridge, Mass.
Buxton, L. (1981) *Do you Panic about Maths?*, Heinemann, London.
Clements, M.A. (1980) 'Analysing Children's Errors on Written Mathematical Tasks', *Educational Studies in Mathematics*, *11*, 1–21.
Collis, K.F. (1975) *A Study of Concrete and Formal Operations in School Mathematics: A Piagetian Viewpoint*, Australian Council for Educational Research, Melbourne.
Dienes, Z.P. (1973) *The Six Stages in the Process of Learning Mathematics*, NFER, Windsor.
Flavell, J.H. (1963) *The Developmental Psychology of Jean Piaget*, Van Nostrand Reinhold, London and Princeton.
Hart, K.M. (ed.) (1981) *Children's Understanding of Mathematics: 11-16*, John Murray, London.
Hawkins, D. (1973) 'Two Sources of Learning', *Forum*, vol. 16, no. 1, 8–11.
Hebb, D.O. (1955) 'Drives and the CNS (conceptual nervous system)', *Psychological Review*, *62*, 243–54.
Johnson, D. and Myklebust, H. (1967) *Learning Disabilities: Educational Principles and Practices*, Grune and Stratton, New York.
Kosc, L. (1974) 'Developmental Dyscalculia', *Journal of Learning Disabilities*, vol. 7, no. 3, 164–77.
Krutetskii, V.A. (1976) *The Psychology of Mathematical Abilities in Schoolchildren*, Chicago University Press, Chicago and London.
Lovell, K. (1978) 'Concept Development' in G.T. Wain (ed.), *Mathematical Education*, Van Nostrand Reinhold, London and Princeton.
Magne, O. (1978) 'The Psychology of Remedial Mathematics'. *Didakometry (No. 59)*, School of Education, Malmo, Sweden.
Mathematical Association (1980) *Pupils' Projects – Their Use in Secondary School Mathematics*, Mathematical Association, Leicester.
May, J.L. (1980) Under-achievement in Mathematics. Unpublished MEd dissertation, University of Bath.
OU (1980) *Mathematics Across the Curriculum*, Open University Course PME 233.
Piaget, J. (1962) 'The Stages of the Intellectual Development of the Child' in

H. Munsinger (ed.), *Readings in Child Development*, Holt, Rinehart and Winston, New York.
Piaget, J. (1973) 'The Affective Unconscious and the Cognitive Unconscious', *Journal of the American Psychoanalytical Association*, vol. 21, no. 2, 249–61.
Rees, R. (1973) 'Mathematics in Further Education: Difficulties Experienced by Craft and Technician Students', *Brunel Further Education Monograph No. 5*, Hutchinson Educational, London.
Richards, P.N. (1980) 'Dyscalculia – A Worthwhile Concept for Educational Research?', *PMEW paper*, Chelsea College, University of London.
Skemp, R.R. (1971) *The Psychology of Learning Mathematics*, Penguin, London.
Skemp, R.R. (1976) 'Relational Understanding and Instrumental Understanding', *Mathematics Teaching*, 77, 20–6.
Weinstein, M.L. (1978) Dyscalculia: A Psychological and Neurological Approach to Learning Disabilities in Mathematics in Schoolchildren. Doctoral Dissertation, University of Pennsylvania.
White, R.W. (1959) 'Motivation Reconsidered: The Concept of Competence', *Psychological Review*, *66*, 297–333.

5 ORGANISATION AND METHODS

Derek Woodrow

5.1 Introduction

It was a common assertion in the early 1970s that although the content of mathematics at secondary level had been changed by 'Modern Mathematics', the style of teaching had not altered very much. This was contrasted with the primary school situation where it was apparent that changes in method were also very common. The 1970s, however, have seen a basic change in teaching style with the widespread adoption of 'individualised' learning systems at both secondary and primary stages of schooling. It is important to see these changes in the context of an ever-developing system:

1945-50 The development of 'secondary modern' curriculum. Grammar schools still following traditional syllabus with separate courses in arithmetic, algebra and geometry.
1950-55 Establishment of GCE O-level. Development and introduction of 'Jeffery Syllabus' courses with the aspects of mathematics treated in a more integrated manner.
1955-60 Adoption of an integrated style examination with much less formal geometry.
1960-65 Establishment of CSE courses. Development of modern mathematics.
1965-70 Adoption of modern mathematics syllabus.
1970-75 Development of individualised learning programmes.
1975-80 Adoption of individualised learning programmes.
1980-85 Development of micro-computing, common 16+ examinations, . . .

It is interesting to note that the gap between entering the secondary school as a pupil and entering as a teacher is about eleven years. Thus a teacher trained since the war has often missed out one whole stage in the development of mathematics teaching and no doubt this is one reason for the occasional appearance of some uncertainty amongst mathematics teachers.

The traditional teaching of mathematics as independent courses in arithmetic, algebra and geometry often hid clear and equally distinct variations in aims, objectives and methods. The styles of teaching and varying content from repetitive practice, model answers, algorithms and set routines, problems and riders all provided a variety of approach coupled with established pedagogic patterns which teachers could and did follow. The introduction of the integrated syllabuses produced a more uniform approach to mathematics with a stress on short term continuity and the use of specific topics such as 'factors' or 'area' on which to centre the teaching for two or three weeks.

Until 1970 nearly all secondary schools organised their mathematics in ability-grouped classes, so called homogenous groupings. In many cases this was based upon general academic abilities, but methods of 'setting' which attempted to form groups more homogenous in mathematical ability were also employed. Even though it was evident that there were inevitably quite wide variations in pupils abilities and attainments they were taught as a class. This did not imply that there was only one style or method of teaching, nor that individuals were not treated differently! The growth of comprehensive schooling led naturally to a strengthening of the social aims and objectives of education to balance academic aims. This has developed alongside a commitment to a developmental, individualistic view of the psychology of learning (epitomised by the work of Piaget, Bruner, Dienes and Skemp) as opposed to the earlier general psychology. One of the most frequently used words in present-day mathematics education is 'concept' and thus much attention is paid to how an *individual* develops a concept. This background suggests some possible theoretical basis for the move to individualised learning. It is not often, however, that educational reasons alone cause major changes, unless accompanied with economic advantage and pragmatic 'solutions' to practical problems.

Two economic factors were present during the 1970s to encourage the growth of individualised learning systems. The first was the shortage (still a serious concern) of qualified mathematics teachers. This led to the idea of a 'teacher-proof' method of instruction which would enable a pupil to develop mathematically even though the teacher could not provide much personal help. The initial concept of an individualised system saw the teacher largely as a manager of resources (see Banks, 1971). Whilst this is not a factor stressed by current schemes it is this feature of relieving the teacher of at least some of the burden which is an attractive and important element in their acceptance by teachers. (The consequences of this are discussed below). The second factor is a directly economic reason in that the cost of replacing an existing series of texts held in a school for a preferred set of texts exceeded the capitation normally allowed to a school. The capitation per pupil for most school mathematics departments is of the order of £1.00 which must provide exercise books, textbooks and equipment. (A typical textbook costs £1.60.) By purchasing a set of work-cards however, costs can be reduced although the organisation can be more complicated if sufficient copies are not available.

Another reason for the changes in method was a direct response to the very great problems raised by moves to mixed-ability teaching. Although there is considerable disagreement on the virtues for mathematics of this organisational change, it is now a common practice for the first two years in many secondary schools. Whether the teacher is in agreement, neutral or opposed to mixed-ability grouping does not alter the major questions it brings to the surface. The particular difficulties which this has caused in mathematics teaching has meant changes not only in methods but also in objectives and the meaning of teaching mathematics which are of wider significance than particular management arrangements.

The reasons for the adoption of mixed-ability grouping in mathematics teaching can be either the overall plan (or insistence) of the school or a decision of an individual teacher. In practice, it has frequently been the school which has made the decision, but regardless of who initiates it the background to the decision is likely to concern social philosophy. One consequence is that the mathematics curriculum should (must?) echo these primarily social aims. As a broad generalisation this philosophy does not seek to insist that all children are the same, but that the similarities are greater than the differences and that differences in various areas of study are often complementary. There is also, above all, a desire for children to respect each other as individuals regardless of academic or any other proficiencies or deficiencies.

The resistance of many mathematics teachers to mixed-ability grouping lies in the history of mathematics teaching in that it is traditionally a goal-oriented subject which is easily and 'accurately' (even if momentarily) assessed. A goal-oriented subject means that it is the end product (answer) that really matters rather than the doing of the task. This approach emphasises the graduation between pupils and makes grouping by ability a natural structure. An approach which is goal-oriented is to some extent contradictory to the philosophy underlining mixed-ability grouping. Any restructuring of mathematics classes into mixed-ability classes will mean it must either change the style of mathematics teaching or it must attempt the difficult task of fitting a goal-oriented approach to a nearly alien philosophy. Both of these are likely to create rather difficult situations before solutions are found.

At the pre-secondary level the size of many classes has meant that mixed-ability grouping has been economically essential. The problems of goal-orientation have been largely hidden at this level for a number of reasons. First, the essential egocentricity of children at this age makes divergence of ability less overtly disturbing to the children, although the latent effects may well be enormous. Secondly, the differences in achievement are cumulative, indeed it could be argued that ability is exponential in growth since each stage is dependent on the previous achievement stages. This will result in an apparently very sudden growth in the divergence of abilities. Thirdly, it has been traditional to depress the achievement of children with greater ability by practising already almost perfect skills. Despite all this camouflage there has been an increasing awareness of the problems attached to mixed-ability classes and considerable changes have taken place in order to try to minimise the effects. Many changes in the teaching of mathematics at this level have been largely, if not quite entirely, aimed at changing to task-orientation. This is clearly seen in the adoption of Piaget's psychology as a rationale for the primary school, and was inherent in the very influential work of both Miss Biggs and the Nuffield Mathematics Project.

This change to considering activity rather than output, produced some misunderstanding from the more goal-oriented secondary schools, which were slower to change in this direction. The reluctance of the secondary schools to change from goals to tasks is not really surprising, since the most prestigious teachers of mathematics at this level (and consequently the usual 'models') have been successful goal-seekers in mathematics.

5.2 Knowing and Doing Mathematics

The eminence of mathematics teaching as a goal-oriented subject is clearly evident. Whilst few subjects are short of goals, none are so short of tasks as mathematics. (It is perhaps this feature which creates such a strange reaction by adults to mathematics, in that there is usually a protestation of fear but very high sales of books on mathematics for the layman.) In almost all other subjects it is clearly possible to involve many children in the same task, even to the extent of asking identically worded questions. The results of the children's work will show considerable variety, but all will be able to attempt what is superficially the same task. In mathematics this is traditionally not attempted – questions can only be done or not done. Indeed, it is frequently the case that questions are not even capable of being done badly! The introduction of open-ended questions and problem situations has shown that this need not be the case, although not many such situations are generally known and still fewer are acceptable as 'real mathematics' by teachers who are not used to working in this way.

One good example is the use of 'arithmogons', which were first described by McIntosh and Quadling (1975). In this problem the sum of the numbers of the circles is placed in the squares.

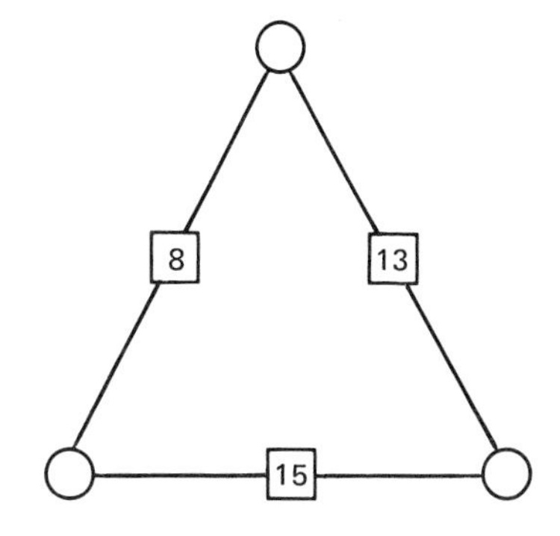

The question is then asked what numbers belong in the circles when numbers are placed in the squares.

Try 8, 13, 15
8, 14, 16
8, 15, 17
8, 16, 17.

By trying different values the pupils 'learn' what makes a good guess at an answer and how to adjust an incorrect guess. If left to work at the problems they develop quite interesting strategies. It also provides them with a real experience of the way numbers balance and check. Various questions arise – under what circumstances do you need negative numbers (and what has this to do with the triangle inequality) – what about a diagram in the shape of a square – why does the square diagram not have one, and only one, solution?

This problem can be solved after some experimentation by most early-secondary-school pupils. If replaced by the more usual mathematical question:

$$\text{Solve} \quad \begin{aligned} x + y &= 8 \\ y + z &= 13 \\ x + z &= 15 \end{aligned}$$

not many pupils would have any idea how to even start solving this particular problem, never mind finding general strategies. Indeed, teaching 'simultaneous equations' has never been an easy task.

The philosophy of goal-oriented mathematics came under some attack as a result of the more general move towards comprehensive education. Often for the first time the academic mathematician faced at first hand the responsibility of achieving goals with the whole school population rather than with a selected elite. It is evident that many of the goals are generally unattainable. It is clearly questionable whether these goals are at all relevant to the average pupil. They are often goals which are required by the professional mathematician, but it is rarely true that all pupils are expected, for example, to become novelists or professional musicians.

The teaching of English and Music, like many other subjects, contain elements that are less evident in mathematics teaching in that they involve as a primary aim an appreciation of the subject. How often are pupils presented with a superb piece of mathematics in the way that they hear, see and listen to the products of other disciplines? It is argued that the appreciation of mathematics implies some minimum competence in the subject, but what this minimum level is has never been established and the natural enjoyment of puzzles and problems would cast some doubt on this as being as important as simply having experienced mathematical activity. This recognition of mathematics as an activity is frequently missing in school curricula, if not in most mathematicians themselves. Activities such as investigating possible nets for cubes, or inventing tessellations, would often not be classified as mathematics until theorems have been discovered. Once theorems regarding these situations have been discovered these theorems become 'mathematics' and are then divorced from the activity which produced them. It is in this way that mathematics has developed to become an almost totally goal-oriented subject.

As an example of the way 'mathematics' is extracted from 'mathematical activity' consider the following account of a discovery:

> The Fibonacci sequence is one in which each term is found by adding the previous two terms. Starting with 1 and 1 this gives, 1, 1, 2, 3, 5, 8, . . .

Dividing each term by the one before gives a sequence which tends to the so called Golden Ratio

$\frac{1}{1} = 1 \quad \frac{2}{1} = 2 \quad \frac{3}{2} = 1.5 \quad \frac{5}{3} = 1.66$

$\frac{8}{5} = 1.6 \quad \frac{13}{8} = 1.625 \quad \frac{21}{13} = 1.615$

and so on → 1.61 8034 . . .

These sequences turn up in many life science and artistic situations (see for example Land, 1960).

Suppose the sequence does not start with 1 and 1?

e.g. 4, 7, 11, 18, 29 . . .
or 3, 4, 7, 11, 18, 29 . . .
or 5, 7, 12, 19, 31 . . .

Then a number of questions might arise, such as what happens if they are continued to the left? The particular question chosen is to consider sequences which contain 37.

. . . . ○ △ □ 37 . . . .

How many are there? What assumptions are needed? One way to make this a sensible manageable question is to insist that the sequence increases (no number to the left is larger) and that negatives are not allowed so that

1 36 37
5 16 21 37

are allowed but not

35 1 36 37
or −6 11 5 16 21 37

With this rule there are 38 possible numbers for □, but 19 of these would not allow a number in △.

How many numbers in □ allow a number to be placed in position ○? Finally, a most interesting question: What number immediately before 37 allows the most numbers to be placed before 37 in the sequence?

In fact the answer is 23 since we can obtain:

			5,	16,	21,	37
			7,	15,	22,	37
1,	4,	5,	9,	14,	23,	37
		2,	11,	13,	24,	37
				12,	25,	37

(There is a need to prove there is only *one* longest. Note the number patterns which emerge in the columns.)

What number (for the same rules) is the 'best' number to place in front of 57? 123? 294?

By testing each of these and gathering other examples it becomes apparent that the 'best' number is nearly $^2/_3$ of the given number. A better approximation would be $^3/_5$ or a better one $^5/_8$ and a better one $^8/_{13}$ . . .

In fact the ratio of the box number to the given number should be as near as possible to the Golden Ratio. So we have arrived at a theorem which leads to a more general theorem of which this is a rider, namely, that the Fibonacci type series starting from any two numbers will have a 'ratio limit' convergent to the Golden Ratio. (You might like to prove this.) For use with pupils below the sixth form this clearly goes too far but a great deal of valuable work on the way numbers behave can be obtained if they can find the useful 'ratio method' for finding approximate answers by using $^3/_5$ or $^5/_8$ and then searching.

This activity provides a good example of one regular pattern of doing mathematics:

Play	– in this case with a paper and pencil.
Questions	– interesting ideas.
Answers	– interesting developments.
Why?	– mature reflection.
Theorems and Proofs	– insertion into official knowledge.

One objection to developing a task-oriented approach is the possible resulting lack of skills and knowledge. There is, however, neither agreement nor evidence of what skills are really appropriate and necessary to the 'average' citizen. This will not be pursued here in any detail but in relation to the general argument about task versus goal-oriented procedures it is interesting to consider the development of skills in regard to sport. In sport the skills, and frequently a great deal of factual knowledge, follows almost entirely upon the pleasure of participation and of spectating, of which more will be said in relation to mathematics later. The activity of collecting 'cigarette-cards' or aircraft numbers brings with it a great deal of information, but it is the activity which is central. In sport, in collecting or even reading books, it is activity without immediate academic or productive objectives which is the initiator and with luck these stimulate the desire for expertise and knowledge which is central to any education. Just as the wish to be better at a particular activity leads to practice and training there is no reason why we should not expect the desire to be good at doing mathematics to lead to skill and knowledge. This is not really a justifiable argument, of course, since the public views or attitudes towards sport and mathematics are very different, and consequently the pupil will respond differently. The ability of pupils to gather information and skills incidentally, however, is highly developed and rarely considered in the school mathematics environment.

5.3 Mixed Ability Groups versus Homogenous Groups

Before looking at different ways of working with children, whether they are grouped on ability or grouped randomly, it is useful to review some of the issues which are involved in these choices. There are, of course, a number of alternatives to the two extremes of streaming (usually based on general intellectual ability rather than on mathematical ability) and mixed ability groups (with grouping based on a supposed random variable such as an alphabetic ordering). It is not unusual to 'top and tail' the ability range by separating out one group of pupils who are particularly good at mathematics and another group who are particularly poor. It is difficult to do this unless the whole year-group (or sometimes a large half-year-group) is block timetabled. Taking the group with the highest general ability will not have the same result as separating on specifically mathematical ability, and many 'remedial' groups are separated on reading rather than mathematical abilities. A fairly large top and tail may mean, of course, that the spread of ability in the remainder might be small since the distribution is likely to be normal. This major group of pupils may be randomly grouped, or split into broad overlapping ability groups or into sets reputedly differentiated into reasonably homogenous groups. Flynn (1972) showed these choices well with diagrams:

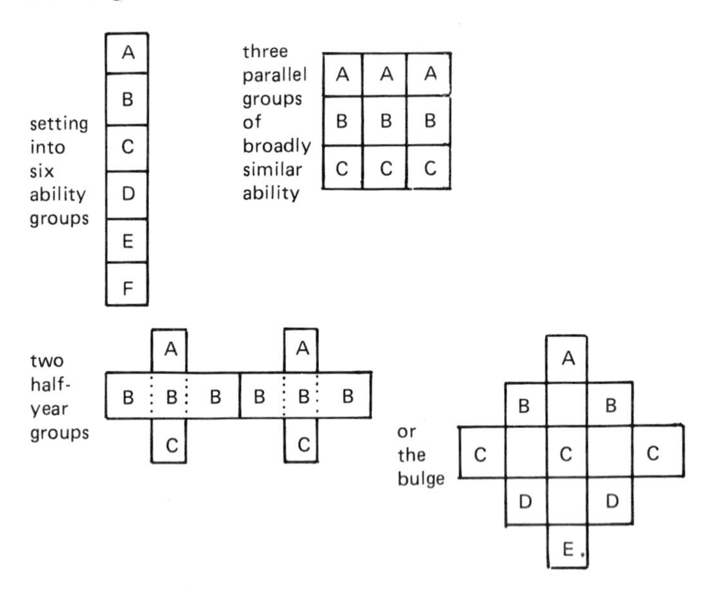

Another consequence of simultaneous timetabling is the opportunity to vary staff commitments. By timetabling, say, six teachers to five groups, a number of variations are possible. The extra teacher can be used, at either end of the ability range, to give more personal tuition, or with different groups as topics with more intensive teacher involvement are tackled, or to move through the groups with an individual monitoring programme to provide more feedback to the group of teachers, or to prepare leading lessons to give to the whole group as a central stimulus.

In considering how the classes should be organised it is important to recognise that academic ability is not the only factor that makes children compatible. Personality, friendships, interest and emotions are also important and must be considered if the optimum working environment is to be provided. It is also clear that children develop at different rates and so could well expect to change ability groups, yet there is ample evidence that once grouped it is rare for changes to be made. This may be partially due to studying slightly differing content in different groups, but there is also good evidence that the children perform to expected rather than possible standards. Once placed in a particular ability group they tend to-conform to its expectations. Even the factors on which mathematical ability is judged change as the secondary school curriculum develops. Arithmetical ability often dominates early on, with logical thinking predominating later and there need not be a high correlation between these factors. Yet the number of 'late developers' is still small and they are frequently viewed with some uncertainty.

Other problems which are emphasised by the debate on organisation are the effects on individual children of any chosen grouping. The presence of children of greater ability may stimulate pupils to greater efforts and to raise their sights, it may on the other hand produce a refusal to compete in the unequal task and a pupil may settle for avoiding attention and doing just sufficient work to satisfy since he cannot excel. The able child may also depress his performance so as not to be too different from his peers despite (or even because of) extra attention and pressure from the teacher. Equally the lower-ability pupil can disturb the class situation, making excessive demands on the teacher's time or creating a nuisance whilst awaiting yet another helping hand. Grouping lower-ability pupils into one set demonstrably commits them to forming a 'sink' which will remain in stable existence until the end of the compulsory schooling period. Attitudes harden and expectations drop and the pupils are often alienated and thus make less progress than ought to be achieved. It is important that the teacher be aware of such situations in order to attempt to alleviate their major effects and also to protect the majority from over-demanding attention to the extremes which can easily develop.

These problems are not solved by the administrative choices which are made, but it is important that teachers periodically review their attitudes. Whether homogenous or heterogeneous groupings are chosen, the onus for alleviating the problems and creating a comfortable, encouraging, stimulating and effective learning environment lie with the teacher and the methods he chooses.

5.4 Teaching the Whole Class

Teaching Specific Topics

Until recently it was normal for the whole class to be considered as a unit, in that all the pupils would be concerned with essentially the same topic and core material. Proponents of individualised learning often unjustly caricature the method as teaching to the average pupil of the class and ignoring the other 95 per cent. This was, and is, rarely the case. The standard procedure of presenting two or three model examples followed by four or five questions, a survey of these questions and then a further ten for homework is often a much more subtle programme than it appears. The teacher, knowing the class, will ask a variety of oral questions during the introduction aimed at a variety of different groups in the class. He is also aware that the model examples and his comments on them will stimulate quite different interpretations by different pupils. Whilst the pupils are trying the five test questions the teacher will often choose to talk to and support the weaker pupils first and then discuss individually the wider issues and generalisations with the higher-ability pupils. The survey by the teacher to the class of these questions will re-inforce the techniques and algorithms and carefully avoid any over-complicated issues which could arise. Many students will work slowly and with difficulty on the test questions, for as Skemp (1971) indicated 'concepts of a higher order than those which a person already has cannot be communicated to him by definition but only by arranging for him to encounter a suitable collection of examples'. These questions help the pupil to explore the points being made in the lesson and play the same role to the secondary pupil that play and experiment do to younger children.

The reliance that this procedure places upon the teacher's intuition and experience is a weakness when the teacher is deficient in these qualities. With the uncertainties which abounded in the change to 'modern' syllabuses many teachers were clearly unable to produce easily the correct variety of question or to delineate clearly to the pupils the techniques and algorithms underlying the mathematics being taught.

This procedure is not, of course, the only one and is only appropriate for certain mathematical developments such as learning specific algorithms or solving specific types of problems. The attempt to teach the underlying structures of mathematics, epitomised by group properties and the generalised use of functions, is a much longer-term strategy. By concentrating on how these concepts can be conveyed to children – essential for the real appreciation of mathematics – attention can be distracted from the short-term programmes needed by children to measure their achievements. There has been a growth of both longer and exploratory-style questions designed to teach children the skills of doing mathematics (solving problems) and of using three or four sets of examples to achieve long-term aims. The pupils do not always have the patience or the ability to retain ideas long enough to provide the necessary patterns. It is important that work should have both long-term objectives of which the teacher is aware and clear short-term achievements to reinforce the pupils learning. Much of current mathematics teaching in secondary schools lacks an identifiable shape for the pupil but consists of working steadily through a continuing treadmill of examples which the teacher rarely gathers into a climax. (see Brissenden, 1980)

Central Stimuli

Whole class (i.e. a large group) situations can be utilised in a quite different way to present pupils with a stimulus from which to move into their own individual or small group development. This strategy is frequently used in the 'humanities' in which a particularly interesting situation is presented to the pupils who then pursue this themselves along the lines of development in which they themselves are moved. At the primary school level the same technique is often used for the whole curriculum, and a topic is presented to the children which is known to be rich in possible developments, and then those interests which arise are followed. One might, for example, investigate the idea of farms, stimulated by a visit or a story or pictures. The topic is then developed to stimulate work in geography, history, or any other area of the curriculum. This is essentially a move from a particular stimulus to more general considerations. Alternatively one might move from a more general stimulus such as 'war', from which will arise particular and more specific interests (such as particular battles), or poetry, or weaponry.

In mathematics both types of presentation might be used. A particular situation such as the number of squares which may be formed on a nine-pin geoboard, or unicursal curves, or tessellating particular polygons can be investigated and the problem then generalised and investigated by the pupils. In the presentation of more general stimuli the teacher could present a wider-ranging introduction to a particular piece of mathematics, from which the pupils could select that part which most attracted them to be researched and investigated. This latter is somewhat akin to the act of listening to a piece of music, or reading a number of poems or looking at well-designed and constructed furniture. The 'spectating' of mathematics is closely related to the idea of mathematical appreciation, an aspect of which has been largely missing from mathematics though television programmes have occasionally gone some way towards this end. It is true that 'model answers' have been a constant feature of the teaching method, but these are rarely (ever?) stimulating and imply reproduction rather than creation. It is also important to note that whereas the first situation described, moving from the particular to the general, implied stimulus to immediate action, other types of central stimulus need not necessarily be intended to imply immediate action for the pupil.

Many of the best traditional mathematics teachers used to be 'carried away' into performing some mathematics or into a general discussion about the value and variety of applications of a topic. This kind of activity is in its essence a social phenomena and needs the stimulus of an audience, a group interacting and responding as a group rather than as individuals. Work on arithmogons can be set up well by such a group interaction to identify the kind of rules to be investigated; as can the Fibonacci topic. The Fibonacci series also illustrates the kind of topic which has a very rich background and the teacher can show how it arises in bees, trees, buildings and even committee selection procedures! With occasional combination of classes into even larger groups the resources in terms of the Overhead Projector and other stimulating presentation materials can be justified. Indeed, if all teachers had two such presentations for each age group, a regular programme would not then be an unduly large burden on a mathematics department.

5.5 Class Groups

One of the most common variations in organisation in primary schools is the use of groups within a class. This enables a compromise between having a very wide range of ability doing the same work and the loneliness of individualised systems. It can, in fact, be more subtle than merely reducing the ability spread since by judiciously choosing the different activities given to each group the demands of the teacher can be much reduced. Practice of skills can often be organised so as not to need too much teacher interaction, as can some investigation situations (e.g. work on geoboards or tessellations or number sequences) once they have been established. Introducing new concepts, however, can need a great deal of verbal interaction with the teacher. By having, say, five groups they can be organised so as to have three groups on self-generating activities, leaving the teacher free to conduct conversation with two groups. Since the number of pupils per group is fairly small the conversation can be of very high quality in terms of direct interaction with the individual. The size of the groups also allows some pupil-pupil interaction at a meaningful level, especially in situations where the pupils are given a common project to develop as a group enterprise. Some primary schools use a variety of subjects, e.g. creative writing, art, science, local history, at one time, so as to ensure a mixture of demands on the teacher. It is equally important that the mathematics teacher in the secondary classroom orchestrates the patterns of working of the groups. It is essential that the styles of working be clearly established so that the groups know *exactly* what they are to do without constant direction. Systems of group checking of answers and group 'games' can provide standard situations for practising and extending skills. A structure for reporting group activities to the class needs to become accepted so that the group can take on this responsibility. As with any complex system the teacher also needs to establish 'holding' activities to which a group can be directed pending either further thought by the teacher or even just to use time sensibly until the teacher is free to concentrate on the group.

The use of class groups will also often follow the introduction of an idea or activity to the whole class. This can enable a group of more able pupils to be given enrichment activities whilst most pupils are jointly practising new skills and ideas and the teacher can concentrate on a group of less able pupils. Although this assumes that the teacher *deliberately* operates an ability criteria in choosing the groups this may not be the case. If pupils are allowed to associate freely in groups they will very often arrange themselves in like-ability groups without any positive movement by the teacher. It is very rare for a bright child to choose to work with a slow learner. Such flexible grouping also enables the pupil to change groups as his ability varies with time or topic. It is possibly important that the teacher does allow (or even engineers) periodic changes in the constitution of groups to enable re-alignments to take place. There are, of course, situations in which the teacher must interfere to prevent some groupings – a pupil deliberately choosing a less demanding group or pupils who are disruptive when in conjunction.

Class grouping was a more common system of teaching in the early 1970s, and the major project of that era, the Maths for the Majority Project, often assumed this organisation. There would be positive gain in its re-introduction alongside class and individual learning systems.

5.6 Individualised (Independent) Learning Systems

The last decade has seen the growth of work-card or work-sheet learning systems on a massive scale as one method of managing the difficulty of a mixed-ability class. One effect of work-cards is usually to break down the traditional rows of desks into smaller groups, sometimes even of single pupils. This organisational change is valuable to a teacher wishing to revise his techniques, since the advent of novel stimuli gives a greater chance of avoiding an habitual reaction and the opportunity to re-rationalise one's handling of teacher-pupil interactions. This interaction is also changed by the actual use of card instruction, since it absolves the teacher from a certain amount of direct instruction and enables freer discussion of what the instructions actually mean, since the cards form a more impersonal instructor. It also helps place more responsibility for working on the pupil himself and thus tends towards encouraging more personal involvement as well as freedom for the pupil to opt out of the situation. At their best work-cards provide the necessary clues for pupils to pursue self-productively a topic with the minimum interference from third parties (the teacher). At their worst they are impersonal, unstimulating sets of procedures to be followed.

Many of these systems do not differ significantly from text-book material, but even then there are clear advantages of such an organisational pattern. Since they tend to be more self-contained and programmed it is easier for the children to progress independently and with less teacher assistance. As a consequence the children can work more easily at their own pace and the differences between children are not so obvious. A further advantage is that they allow for the insertion of more open-ended activities and the insertion of special programmes for both the clever and slower children.

There is a tendency, however, for these advantages to be somewhat dissipated. The structured programme needed to achieve immediate coherence for a small number (often one) of cards makes it difficult to achieve overall coherence for the course. The latter leads to structure between cards and provides a real means of competitive comparison by the children. Indeed the extent of this structuring often makes the word-cards indistinguishable from a text-book except for the absence of bindings. The use of extra programmes for slower children then tends to stigmatise them and creates failures of confidence. There is also a tendency to seek the kind of perfection in each card with a search for the best method of achieving a particular objective – ignoring the many differences between the children, teacher and classrooms which supposedly justify many of the systems. The insertion of extra programmes to suit each individual child is not a realistic or achievable objective. Indeed most individualised learning programmes are nothing of the kind. They are generally individually *paced* programmes in which all the pupils pursue basically the same series of activities at their own pace. It is often in order to achieve economy that the order of these tasks varies with a permutation of, say, 20 short tasks amongst a group of children. Where a pupil is clearly not achieving on a course or is clearly finding the work very easy then special arrangements are made, this is usually true in any classroom organisation and is not a characteristic solely of work-card systems. It was for this reason that the Schools Council (1979) suggested that such systems should be called *independent* learning systems rather than *individualised* learning systems.

The strength of these criticisms depends to a large extent on the ability of the teacher to manage the situation. Where a teacher has created the set of cards and knows them thoroughly he has the flexibility to adjust and amend intuitively with a security that he can guide the pupil back onto the general pathway. Where a teacher is using a large collection of other people's cards – commercially produced or not – then he is reluctant to move away from the main path since the return route is never clear. Indeed some systems are so complicated and large and the task of merely organising and managing them is so enormous that there can be no question of adaptation or manipulating the system. This is indeed one of the major ironies, that the greater the flexibility and alternatives a system offers the more difficult it is to control and the teacher has to be more rigid in his organisation in order to cope. (It is the problem of having too many sweets to choose in a sweet shop which leads people to choosing the same ones every time!) Although self-construction of work-cards does enable the teacher to know them more thoroughly there is still a problem since:

> The belief that any teacher can devise a workcard of attractive appearance and interest, that uses appropriate language, that caters for a wide spectrum of ability, interest and aptitude (as well as being valid in terms of subject content) has led in many schools to the production of mounds of badly designed, dull worksheets which do little other than include boredom. (Schools Council, 1979)

One essential feature of beginning a work-card system is to choose initially only one or two groups of children and fairly small collections of cards. These cards can then be added to or replaced gradually in a manner which allows the teacher to control and manipulate the organisation, and to be fully conversant with the work required of the pupils. Most card systems can enable some containment if the number of cards in the system is limited by repetition. It is now common practice to give children a list of nine or ten cards to consider in a non-specific order followed by a test of some kind. If these are considered in say, three trios of cards then one set will satisfy three children, a class of 30 children would need 90 cards. If this comprised ten different groups of nine cards, the work could be expected to last about 20 weeks, or half the year. If, however, the test at the end of each group of cards is to be effective it must be acted upon and the teacher must be able to respond to the results. This will lead to the insertion of additional practice cards or the deletion of unnecessary repetition, and this implies the existence of yet more cards. The process can also be made more manageable by the insertion of longer term activities such as pupils' own projects to which they can be diverted to slow down the passage through the card system and give the teacher breathing space. There are other very good reasons for using pupil projects other than easing the teacher's burden, but this is one valid argument for their use.

The length of the list of cards given to a pupil depends upon the kind of tasks which are demanded on the cards (ten minute activities or whole week investigations) and the speed at which a pupil can work. Since it is assumed that each pupil has a mixed profile, they may be good at some aspects but poor at others, the time needed can at best be only an approximate assumption for each pupil. It is important, however, that a pupil should be brought to a teacher's attention to be considered and re-assessed at fairly regular and frequent intervals. A period of two weeks is often aimed at, but for 30 pupils this will mean a new assessment for three pupils every lesson and if the review is taken seriously this could be very demanding when taken alongside the other interventions which the teacher needs to make. Indeed this often leads to the setting of fairly stereotyped lists of cards. These lists are sometimes called *matrices* but since this term has such a specific and universally accepted different meaning in mathematics the adoption of this term would be unfortunate. Ideally the timing and regularity of teacher interference would vary with the personality of the child, some like constant reassurance whilst others prefer to work in a more isolated manner (in any case should these preferences necessarily be encouraged?). Such situations cannot be prescribed but a good teacher knows his pupils and he will often instinctively adapt and amend his programme to go some way to meet these variations. We know far too little about the problem to move from this intuitive level, and to provide inexperienced teachers with some guidelines.

Another way of relieving the teacher of pressure, although it reduces teacher/pupil interaction, is the use of self-checking cards. Essential teacher intervention can be retained, however, by ensuring that with such cards any major failure (e.g. less than 80 per cent correct) is reported to the teacher. These sets of cards can be difficult to create but can lead the pupil to confirmation of a concept without too intensive an involvement by the teacher. Equally useful, although again complicated to produce, are situations in which a pupil having marked his own work is directed by his wrong answers to specific correction-programmes. This was the intention behind much of the work on programmed learning in the 1960s, and produced elaborate branching programme techniques. These did not become popular for a variety of reasons, one of which was a belief that the reasons for errors are so varied that the remedial programmes needed too much flexibility to be practical. Whether it is really desirable to so minimise teacher involvement will be touched on later, but the added complications are very great and it is doubtful if we have enough knowledge and skill at setting up such learning situations to be sure of success.

Self-marking by pupils has always been a feature of teaching mathematics. Not only does it relieve the teacher of yet another task, releasing time for planning and direct involvement in lessons, it also emphasises to the pupil his own responsibility in seeking accuracy and understanding. (It is said that an English pupil believes that he fails because he was badly taught whereas a Scots pupil believes he fails because he did not learn the work properly.) It is often lamented that teachers do not go through the task of marking a whole set of books. With many classes with a large number of pupils, however, the task can often only be achieved in the time available by a mechanical comparison of answers rather than the imagined activity of reading through all the pupils' work line by line.

Pupils often misread a question rather than not actually perform the task required. For example 'Imagine you stand at the centre of a compass pointing north; what direction do you point after turning 90°, 180°, 250° . . .' is presumably meant to test an understanding of compass points and angle measurement. If the child stands *outside* the compass looking at it rather than pretending to be the pin on which the hand is suspended then the answers will be wrong even though his mathematical knowledge is used correctly. Without investigation the answers marked wrong would suggest mathematical incompetence rather than an alternative interpretation. A perceptive, experienced teacher can, it is true, learn a great deal from a rapid scan of the work presented when a large number of pupils are presenting the same work as is normal in class teaching situations. The same repeated error attracts attention and investigation. With individualised systems of working each pupil is presenting a new challenge to the teacher and no patterns of clues are developed to speed the teacher in the checking. Indeed, it may be much more difficult for a new teacher to build up a store of typical children's responses to particular topics when he experiences them in a more random and disjointed sequence. The task of marking work from pupils takes much longer with such variability, and one consequence is probably that a particular pupil's work is read more thoroughly but much less frequently. Some teachers, under pressure, may still glance through work at the same fast pace but without having the same heightened response to the particular problem and this superficial checking is neither constructive to the pupil nor informative to the teacher.

A well-organised assessment pattern, which often accompanies a well-designed set of work-cards, can give advise to the teacher over a period of time. The teacher needs to record not just marks but the way in which a pupil gains those marks if he is to be able to diagnose and develop a pupil's strengths and weaknesses. With a class teaching system the teacher knows the topics which he has taught and can relate a string of marks to the particular topics tested, although how often this is done rather than 'averaging' the marks is debatable. In theory, therefore, skills, processes, memory, problems involving visualisation, learning algorithms, problems of symbolising, the ability to test and prove conjectures could all be extracted from the mark list by analysing the work set. (The APU assessment framework is the start of such an analysis (APU, 1981).) With individualised systems the marks, or results, of a test need to be recorded against the items being tested so as to know what each pupil has or has not tackled. It has become increasingly clear that we need to record what topics a pupil has studied, the level of success in understanding that topic and how well he has retained the work and 'internalised' it into his schemes. The record system needs to contain:

(1) what topic the card was about (e.g. fractions)
(2) what aspect of that topic (equivalent fractions)
(3) the kind of intended reaction by the pupil (e.g. understanding or just an algorithm for finding equivalent fractions)
(4) the level of pupil's responses (can he 'do' or has he 'understood'). This might also include a 'X 0 +' code for 'below par/par/above par' as a response for a particular pupil.

By using a sensible pro-forma which lists the topics and a suitable breakdown of the levels of achievement and/or understanding being considered this can be made reasonably simple to maintain. An item such as

Fraction: Equivalence of $\frac{1}{2}$, $\frac{2}{4}$ .... ☐

Equivalence for $\frac{1}{2}$, $\frac{1}{3}$, $\frac{1}{4}$ .... ☐

Equivalence for $\frac{m}{n}$ .... ☐

with ◪ meaning has met the card

☒ meaning has completed the card

☒ meaning has successfully completed the relevant test

can be quickly and easily maintained by the teacher. Such a wealth of information on each pupil can be overwhelming, and unless the teacher carries out a periodic review it can also be unnecessary. With such a large amount of data it is essential that the teacher has well defined the important elements. Responses to different collections of test items can indicate different problems or strengths of the pupil. The teacher will also need to review the success or otherwise of the particular cards or sheets he is issuing. To determine what is the contribution of the resources or what part is played by the pupil to a particular outcome is not easy but it is an essential feature of the professional activity of the teacher.

The recording system also needs to serve other purposes. Pupils need the reassurance of knowing that the teacher is concerned and involved in their success and failure. Each pupil needs the exhilaration of a period of success in mastering a topic, and even the brightest pupil needs to have deficiencies explored and exposed. The record system should not, however, be so dominant that it interferes and sours the pupil/teacher relationship. It is particularly important for the less able child that the recording does not emphasise and project his weakness but takes into account and expresses the pupil's more positive characteristics. The record system should present the majority of the children with attainable and clear-cut goals. The problem of the mathematics syllabus as a never ending conveyor belt along which children are passed as quickly as possible needs to be countered. Involving the pupils in recording their own development can help give them a stronger feeling of progress and achievement and provide useful motivation.

Records must also serve to help the teacher discharge his responsibility to school and parents in ensuring that each child is developing in mathematics as fully and efficiently as the teacher is able and the system allows. It is often extremely difficult for a teacher to distinguish the very bright, lazy child from the less able, conscientious pupil. The pupil who completes enough work sufficiently well to avoid any interest from the teacher is very difficult to diagnose but may well be far from unusual at the secondary-school level. In adolescence many pupils are often unwilling to be closely scrutinised and attention avoidance is probably as important a feature of character as is attention-seeking by the child who attracts concern for his problems. The record of a pupil's progress needs, perhaps, to be placed against other more random assessments of a pupil and discrepancies sought and pursued. Annual examinations *if properly constructed* can be one source of such a test, but they need to include items directed towards bringing out abilities rather than memory or impressed techniques (see Chapter 10).

Equally valuable is the contrast between homework and classwork. This contrast is difficult to unravel since although the way a pupil works in class can be observed it needs perception to know whether homework is done aided or unaided, with the concentration of being alone or the off-handedness of being unsupervised. The nature of homework within an individualised work-card system is not always well thought out. Work set for homework needs to be capable of being done without teacher support or resources, and since cards are easily mislaid it is frequently constructed outside the class work structure. Ideally homework is for practice of known skills and *not* the learning of new concepts and processes. This again emphasises the importance of the teacher knowing intimately the questions he is asking. The format of setting homework needs more discussion than is possible here.

5.7 The Method in the Message

The chapter began by discussing the differing aims and objectives and their relationship to the methods adopted by the teacher. The various meanings of mathematics and the value attached to them is also conveyed and enhanced by the methods chosen. Some teaching methods are best suited to passing on certain kinds of information or dealing with certain activities. Introducing a topic or an idea which is new to a pupil is difficult to do at the distance imposed by predetermined resources such as the printed page. By presenting a variety of carefully constructed examples it is possible to go some way towards presenting new ideas but can rarely displace the exploratory interchange between teacher and pupil. The introduction of algebraic notation, drawing graphs, transformations, are all more difficult to describe without 'hand waving' and the meaning of processes such as 'proof' need teacher-pupil interaction to crystallise into reality. After the introduction of a topic, however, a carefully constructed exploration of its boundaries can be accomplished without much teacher intervention other than the checking of pupils accuracy and expression. At this level of extending the pupils experience of a concept the work is essentially an individual or small group activity. As Skemp (1971) makes clear, much of mathematics operates at higher levels of concept each of which is built on wide experience of lower-level concepts. The individualised methods now available provide a wide base of lower-level concepts very valuable for developing these higher concepts, but this development needs further teacher intervention and stimulus. Just as the very young child needs the ambiance of hearing adults use sophisticated language and seeing adults reading books to prepare him for later language growth so too do most pupils need such a vision to lead to higher-level mathematical concepts. The language and structures of higher-level concepts must be available when needed and this is one activity the teacher must engage in or the connection may not be established.

It is the constant interchange from group to private activity which makes mathematics into a social activity. Although much activity is performed privately there is a need to return and be restimulated and to restructure ones ideas by comparison with others. This interchange can be with the peer group, and pupil discussion is a valuable ingredient in learning mathematics. Recent HMI reports have stressed, however, the lack of teacher-stimulated mathematics language in the classroom. Complicated individualised structures have led to a preponderantly management style and content of the language:

> Teacher-pupil discussion is often one-sided, seen only in terms of help, instruction, assessment and explanation, and not in terms of encouraging response from the pupil. The teacher needs to establish a relaxed classroom atmosphere so that he has time to

> think on the job, has time to listen to what the pupils are saying (the best method of assessment) and to talk *with* the pupils. (School Council, 1977)

One of the often expressed advantages of individualised learning is the value of the intensity of individual discussion between pupil and teacher. Certainly the choice between high intensity but short individual conversations and lower intensity but longer group discussion is not a simple one. Sometimes width rather than quality is more significant. Where the individual discussion relates largely to management matters then its mathematical value is clearly diminished. The teacher is an essential and crucial factor in the classroom to quote from Brissenden (1973):

> The teacher is a competence model for his pupils not only in matters of style and layout of solutions but also in all the aspects of the manner in which he creates and handles problems. Amongst these aspects are such activities as:
> (1) Constructing questions about situations, generating hypotheses;
> (2) Deciding on appropriate conditions and methods of solution;
> (3) Evaluating one's own choices and those of others, arriving at agreement by some logical decision-process.

(This is further developed in Brissenden, 1979.)

5.8 Some Conclusions

One of the major useful outcomes of the reforms of the 1960s in mathematics teaching has been to diminish the belief in instant solutions. Indeed, solutions are unlikely to be found other than in a slow and steady move towards greater effectiveness. Behind many of the headlines which preceded the 'great debate' and led to the establishment of the Cockcroft Committee has been the implicit assumption that if teachers are only told what to improve they can do so. That is not the way changes in the education system operate. Changes are effected slowly and improvement comes in small increments.

Individualised learning cannot overcome the lack of intuition of the mathematically weak teacher, and a *teacher-proof* curriculum cannot but diminish and impair the standards sought. Individualised systems do, however, give the teacher a splendid alternative resource to achieve many of his objectives and release him from some tasks in order to deal with others which are more dependent upon his intervention and control. The teacher needs to fashion a programme of variety and pattern to attract and maintain a pupil's commitment to work. A careful programme of class lessons (introductions, drawing out communalities, exhibiting mathematics), mixed with group activities (projects on applications or investigations of large-scale problems) and individualised work (practising skills, developing and exploring concepts, exploring a problem) needs to relate content to method and activity to process.

Bibliography

APU (1981) *Mathematical Development: Primary survey report No. 2*, HMSO, London.

Banks, B. (1971) 'The Disaster Kit', *Mathematical Gazette*, *391*, Feb. 1971.

Brissenden, T.H.F. (1973) 'Individualised learning and mixed ability groups in secondary mathematics', *Mathematics in Schools*, vol. 2, no. 6, 16–18.

Brissenden, T.H.F. (1979) 'Teacher-pupil discussion in mathematics', *Mathematics in Schools*, vol. 8, no. 3, 29–31.

Brissenden, T.H.F. (1980) *Mathematics Teaching: Theory in Practice*, Harper & Row, London and New York.

Flynn, F.H. (1972) 'The Case for Homogeneous Sets in Mathematics', *Mathematics in Schools*, vol. 1, no. 2, 9–11.

Land, F. (1960) *The Language of Mathematics*, John Murray, London.

McIntosh, A. & Quadling, D. (1975) 'Arithmogons', *Mathematics Teaching*, *70*, 18–23.

Skemp, R.R. (1971) *The Psychology of Learning Mathematics*, Penguin, London.

Schools Council (1977) *Mixed Ability Teaching in Mathematics*, Evans/Methuen, London and New York.

Schools Council (1979) *Discussion Notes on Curriculum Planning: Mixed-ability Grouping*, Schools Council, London.

6 THE SLOW LEARNER AND THE GIFTED CHILD

Renée Berrill

> There were indications that the ablest pupils were not always sufficiently challenged. There were more widespread indications that the less able commonly had inappropriate programmes, which could be both too difficult in some respects . . . and yet insufficiently demanding in others. (DES, 1979)

Her Majesty's Inspectors, in their report on secondary education, were not referring only to mathematics but, in the supplementary information on mathematics, it was noted that: 'A need to organise new courses for less able pupils was perceived in 68% of all comprehensive schools and in 60% of secondary modern schools' (DES, 1980). In almost three-quarters of these cases this recommendation was seen as 'being a very strong one'. Also, abler pupils 'require a challenge to work in different ways' (DES, 1979) in mathematics.

6.1 Factors Influencing Children's Learning of Mathematics

What are the main factors that influence and encourage children's learning and the learning of mathematics in particular? Amongst psychologists and mathematicians who have attempted to answer this question there is a fair concensus of opinion that, whilst home and parents probably have the greatest influence, motivation and attitudes play a major role. Piaget's research (Inhelder and Piaget, 1966 and others) emphasised how important it is that educationists should recognise that all children develop intellectually in stages. Whilst some details of his research have been criticised and some of his theories questioned, work done subsequently by Lovell (1971), Skemp (1971), Krutetskii (1976), Donaldson (1978) and many others stresses the importance of the gradual development of the child and of his *understanding* of the subject he is learning. Dienes (1960 and 1973) maintained that the environment in which the child finds himself is of the utmost importance for him to develop logical concepts. Skemp also stresses the importance of 'experience' stating that 'everyday concepts come from everyday experiences' (Skemp, 1971). Krutetskii (1976), working in Russia during the years 1955 to 1966, criticised Western psychologists and their testing methods for discovering levels of intelligence and mathematical ability and set out to discover *what* mathematical ability really is. His hypothesis was that 'able pupils may be characterized by a distinct flexibility in their mental processes, . . . by an ability to generalise, . . . and they will be able to think spatially' (Krutetskii, 1976). He assumed that able and less able pupils will differ in their rate of curtailment. This implies that the able child can solve problems without going through all the processes of argument whilst the less able child needs to be guided through the whole argument. Whilst Krutetskii criticised Piaget's 'stage' approach he used Piaget's experimental methods of talking to children on a one-to-one basis.

6.2 Motivation

'Learning is most efficient when motivation is intrinsic' (Biggs, 1967). Biggs explains that what is being learnt must be interesting and there should be some form of extrinsic motivation too, for most children. Hopkinson maintains that above all, most able children are motivated to use their talents in spite of external pressures and 'the roots of motivation seem to lie in an eager curiosity' (Hopkinson, 1978). Unfortunately, the educational system, as it exists today, stifles eager curiosity in both able and less able children. The methods used in the teaching of mathematics are geared to the examination syllabus and text-book, but some of the most interesting and challenging parts of mathematics exist outside the syllabus. The teacher will, so often, not divert from the fixed syllabus, either because he is unsure of himself as a mathematician or is afraid of not getting his class adequately prepared for the examination. Because of this attitude he is often doing more harm than good and is, in fact, insulting the intelligence of many of his pupils at all levels of ability. Motivation is not likely to be intrinsic for the slow learner and it must be recognised that many children have great difficulty learning mathematics and in retaining facts. The teacher, therefore, has the task of motivating these children to want to learn recognising that their biggest problem is 'learning how to learn' (Williams, 1970). The importance of the home background and the attitude of the parents, as already suggested, cannot be emphasised too much. At the pre-school stage it is important that children should communicate with adults, should encounter books and newspapers and should be encouraged to play imaginative games with boxes, containers, sand and water. A lack of these experiences at this early stage may discourage a desire to learn at a later stage. Adverse attitudes from parents and older siblings can produce negative motivation whilst encouragement, help and praise from parents can have a positive effect on a child's attitude to learning. When motivation is not naturally intrinsic someone must produce some form of positive extrinsic motivation.

6.3 Teaching Methods

A relaxed atmosphere in the classroom giving children of all abilities the opportunity to discover mathematical ideas and relationships for themselves is essential. All children should be enjoying and understanding the work they do in the classroom and this is true for the slow learners as well as for our future mathematicians. Research into the attitudes of 13-year-old children towards mathematics indicated that children of all abilities considered that 'enjoyment' of their work was important to them whilst, in general, the slow learners were more concerned about 'understanding' than the more able children. (Berrill, 1980)

The shortage of good mathematics teachers continues to cause concern and, inevitably, this situation affects the way mathematics is taught in school. If a teacher is unsure of his own mathematical ability he will not want to encourage able children to ask questions or to discover for themselves. The result is that 'most of the mathematical processes they learn to carry out do not make any real sense' (Dienes, 1960) and the children are simply learning and reproducing facts in a rote manner. Unfortunately this approach appears to work, for the children pass their examinations.

Recommendations were made by Her Majesty's Inspectors (DES, 1980) that, in many schools, there should be new appointments made of teachers 'with appropriate qualifications'. This is obviously desirable but in an age of recession and with a shortage of qualified mathematicians it may not be possible to implement quickly. The recommendation must still stand, however. A more realistic suggestion was the advisability of more in-service training for heads of departments and assistant teachers of mathematics. The implementation of this suggestion should be treated as urgent if the future of mathematics, technology and science is seen as not only competitive in the world of the future but also as exciting and worthwhile studies in themselves.

6.4 The Slow Learner

Many people have tried to define the slow learner and labels such as 'backward', 'retarded', 'less able', 'dull', etc. have been attached. The Warnock Report (DES, 1978) has suggested that we should stop trying to label children for what they *are* and start thinking, instead, of their needs. However, in order to consider their needs in mathematics we must indicate the children to whom we are referring. A child may be a slow learner for a number of reasons; he may be under-achieving in one subject of the curriculum only or in all subjects. He may just need some extra help in one topic of one subject. When the Schools Council Low Attainers in Mathematics Steering Committee was searching for a definition they decided that they were aiming their deliberations at about the bottom 20 per cent of the ability range bearing in mind that less than 2 per cent of the whole school population is in Special Schools. Berrill (1980), when conducting research on 'attitudes', defined the slow learners as those children who fell into the bottom 20 per cent of the children in normal schools on the results on an objective mathematics test (not on IQ). The National Council of Teachers of Mathematics (1972) in America, recommended that no attempt should be made to identify slow learners beyond the general definition of 'students who are not achieving at the desired level'.

It is interesting to note that when teachers are asked whom they reckon are the slow learners their answers vary from 10 per cent to 60 per cent of all children in school. Many mathematics teachers feel that it is the group of children just above the 'bottom 20 per cent' whom they find the most difficult to teach and it is, therefore, suggested that what is to follow in this chapter may be interpreted and adopted by the teacher for *any* child who needs extra help in mathematics.

The reasons for under-achievement in mathematics are many and varied. Apart from specific handicaps such as cerebral palsy, hearing or sight impairment, the reasons may fall into two broad categories – deficiencies in cognitive functioning and deficiencies in affective functioning, although many pupils may be under-achieving due to no fault of their own. The slow learner may have weaknesses in intellectual skills or he may have adverse attitudes and a poor self image. In fact he is likely to be deficient in both these areas.

Slow learners are concerned that they should be able to enjoy and understand the mathematics that they are learning (Berrill, 1980). Some of the reasons for not understanding are:

(1) they cannot read the text book or worksheet
(2) mathematics is too abstract and apparently irrelevant
(3) there is too much symbolism in mathematics
(4) they are bored with the diet they get in mathematics lessons.

Children are likely to be motivated to learn mathematics if:

(1) they enjoy it
(2) they understand it
(3) they think it is going to be useful
(4) they want to pass an examination
(5) they like the teacher and his/her methods
(6) they can do it and enjoy getting their sums right
(7) gratification is immediate

and they are likely to lack motivation to learn mathematics if:

(1) there is little or no home support
(2) they do not understand the language and symbolism
(3) there is too much abstract work
(4) they do not understand the content
(5) they are bored
(6) it appears to be irrelevant
(7) they are always getting their sums wrong
(8) it is too difficult and they cannot cope
(9) they do not like the teacher and his/her methods
(10) they have reading problems
(11) they have been absent a lot and have fallen behind their peers.

The lack of home support may result in:

(1) lack of sufficient sleep or an inadequate diet leading to retardation in intellectual progress
(2) truancy, thus creating gaps in a child's knowledge and understanding
(3) the development of adverse attitudes towards school generally and often towards mathematics in particular.

This lack of support should be compensated for in the school situation. Lack of sleep or a poor diet are very difficult to combat but, at least, an understanding teacher can help. Truancy should be checked as soon as possible. If school is interesting enough the child will not want to truant but if he does, the reason for such behaviour should be investigated. The development of attitudes is of very great importance and teachers and schools should be encouraging positive attitudes. Mathematics should be seen as a subject to be liked and enjoyed, not as one to be hated and/or tolerated. Since slow learners tend to require immediate gratification continuous feedback and encouragement is likely to help to create positive attitudes.

Mathematics has a language specific to itself but it also uses words which are used in everyday life with a different meaning. 'Vulgar', 'foot', 'field', 'differentiate', 'form' are such words, whilst 'isosceles', 'equation', 'square root', 'trapezium' are words which are more specific to mathematics. Some children have great difficulty in understanding and remembering the meanings of some of the words used in mathematics. How often, one wonders, does a child think he understands what 'subtract' means when someone uses the expression 'minus' or 'the difference between' and he becomes utterly confused? Teachers should be aware of these language problems when introducing new words and ideas to children.

Mathematicians claim that mathematics teaches children to 'think logically' but, unless it is taught with, and for, understanding, there seems to be no reason why children should recognise any logic in it. It is just a collection of magic boxes! Slow learners will continue to get more black crosses than red ticks because they cannot remember which magic box to open at the appropriate time.

Some children have reading problems and cannot, for example, distinguish between b and d or between 'saw' and 'was' or b and p. If this is so they are likely to have problems distinguishing between + and X or $>$ and $<$ or $\times$ and x or

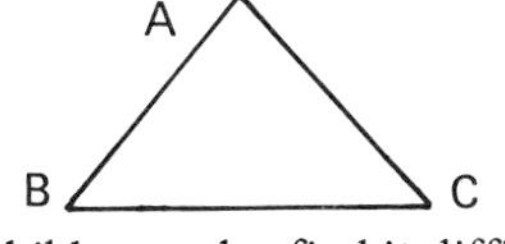

and

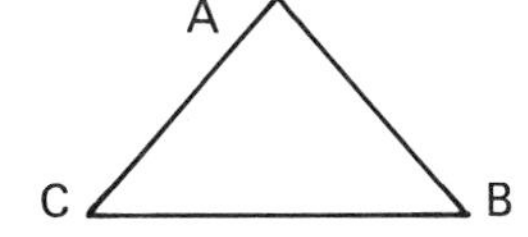

A child may also find it difficult to do a sum such as

$$\begin{array}{r} 23 \\ +45 \\ \hline \\ \hline \end{array}$$

because he has to work from units on the right to tens on the left – a right to left procedure – whereas when he learns to read words he works from left to right. The use of Dienes blocks will probably overcome this problem.

Often, if a child has had difficulty in learning to read, he is quite unable to understand the contents of the mathematics text-book or work-sheet. This will inevitably hamper his progress in learning mathematics. It has been noted that sometimes, when a child is given some mathematical problem to do and no reading is involved, he proves to be quite adept at solving the problem. The use of concrete materials, diagrams and graphs thus become an essential method of approach with such children.

If arithmetic is taught to very young children just by doing sums without recourse to the concrete they are likely either to learn techniques only without understanding or they will not learn at all.

The Teacher and the Classroom

'Too many unhappy children are being taught by unhappy teachers' (Pringle, 1975). This is an indictment on teachers and, whilst Pringle is not referring just to mathematics it is true to say that there are too many children at all age levels who are bored with the mathematics they are being taught. The classroom for slow learners should be bright and cheerful (not the dark corner of a corridor!), there should be colourful charts on the wall and also lots of the children's own work. Equipment should be on view and not tucked away in a cupboard. The teacher should exhibit an enthusiasm for mathematics whilst showing an understanding for children's learning problems.

An ability to test children, to diagnose their difficulties and then to give remedial help is something that all teachers of slow learners in mathematics should be able to do. However this is not always included in the training course of teachers of mathematics and is often left to the remedial teacher.

Pre-number Concepts

Number recognition should begin with children playing with their own toys. They should be encouraged to recognise the 'conservation' of number and the use of building bricks can be helpful (see p. 113).

This concept of conservation should be acquired before progressing to number bonds and arithmetic. Slow learners are likely to take longer to understand and much patience on the part of the teacher is necessary. It is probably true to say that many children fail to

e.g.

(1)

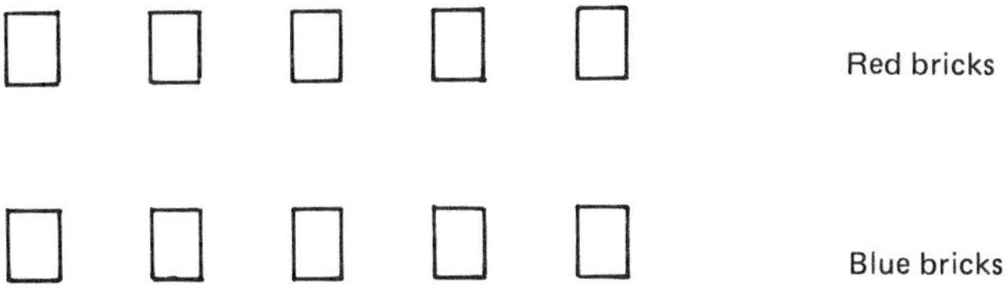

Red bricks

Blue bricks

There is the same number of blue bricks as red bricks.

(2)

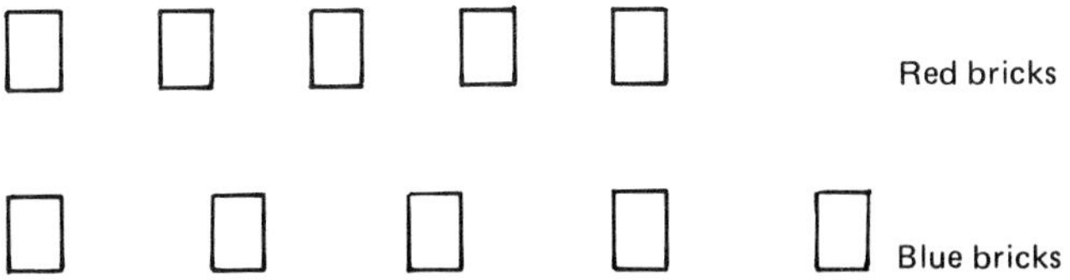

Red bricks

Blue bricks

There is still the same number of blue bricks as red bricks.

(3)

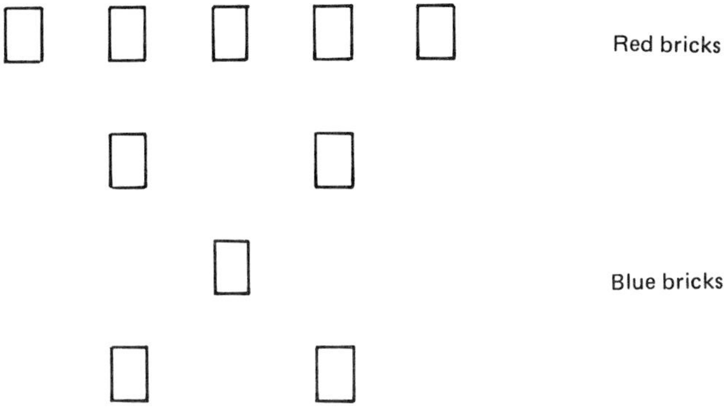

Red bricks

Blue bricks

There is still the same number of blue bricks as red bricks

understand mathematics in junior or secondary school because they are not given enough time and opportunity to acquire these early 'pre-number' concepts.

The idea of one-to-one correspondence now develops and children can be asked the question 'How many?' Again their own toys and everyday articles should be used.

e.g.
(1) If four people are going to sit at a table for a meal, how many plates, cups, saucers, knives, forks etc. will be needed?
(2) If we have three teddy bears and each one must have a bow, how many bows do we need?

Comparison of size and ordinal numbers are important concepts and slow learners need a lot of practice with these ideas.

e.g.
(1)

Jim

John is taller than Jill
Jill is taller than Jim
John is taller than Jim
John is in first place

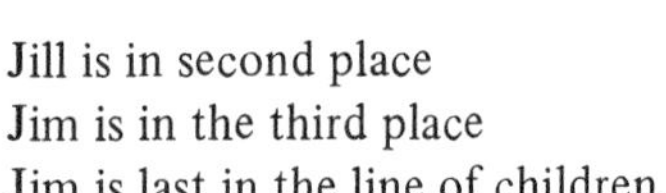
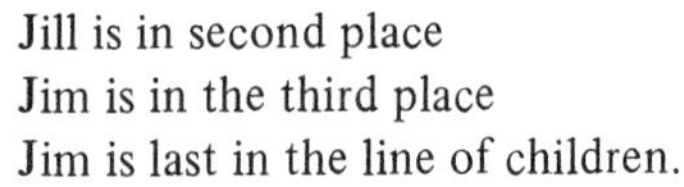

Jill is in second place
Jim is in the third place
Jim is last in the line of children.

Children should be encouraged to use their own toys (e.g. dolls) to gain these concepts.

Cuisenaire rods are also useful for this work. The use of these rods will also lead children into arithmetical ideas such as the four rules of number, fractions and even indices.

Arithmetic

Early work on number bonds and the four rules must be approached in a concrete way and it should not be assumed that older junior or even secondary children do not need the concrete approach. Many children do not reach the 'formal operations' stage when thinking becomes abstract and divorced from real life situations until the age of 15 or 16 years and some slow learners never reach it. However, the aim must be to 'nurse' them carefully through the concrete stage with the hope that they will reach the formal stage at least for some concepts.

Pebbles, conkers, peas, beans or unifix bricks or children's own toys should be used to introduce the concept of a set. A set is a collection of objects.

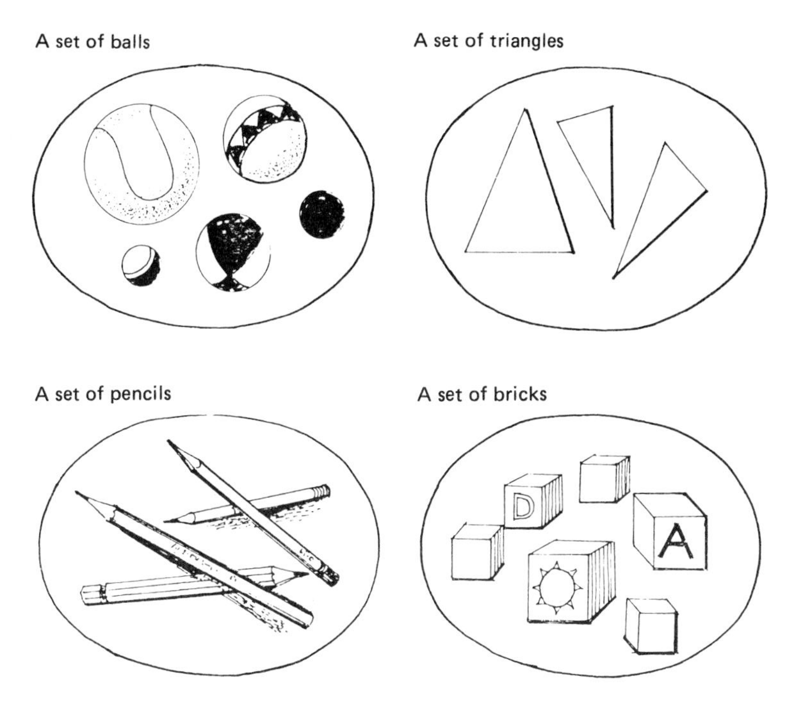

Addition can now be taught but the number symbols should not be introduced until the child has an understanding of the concept.

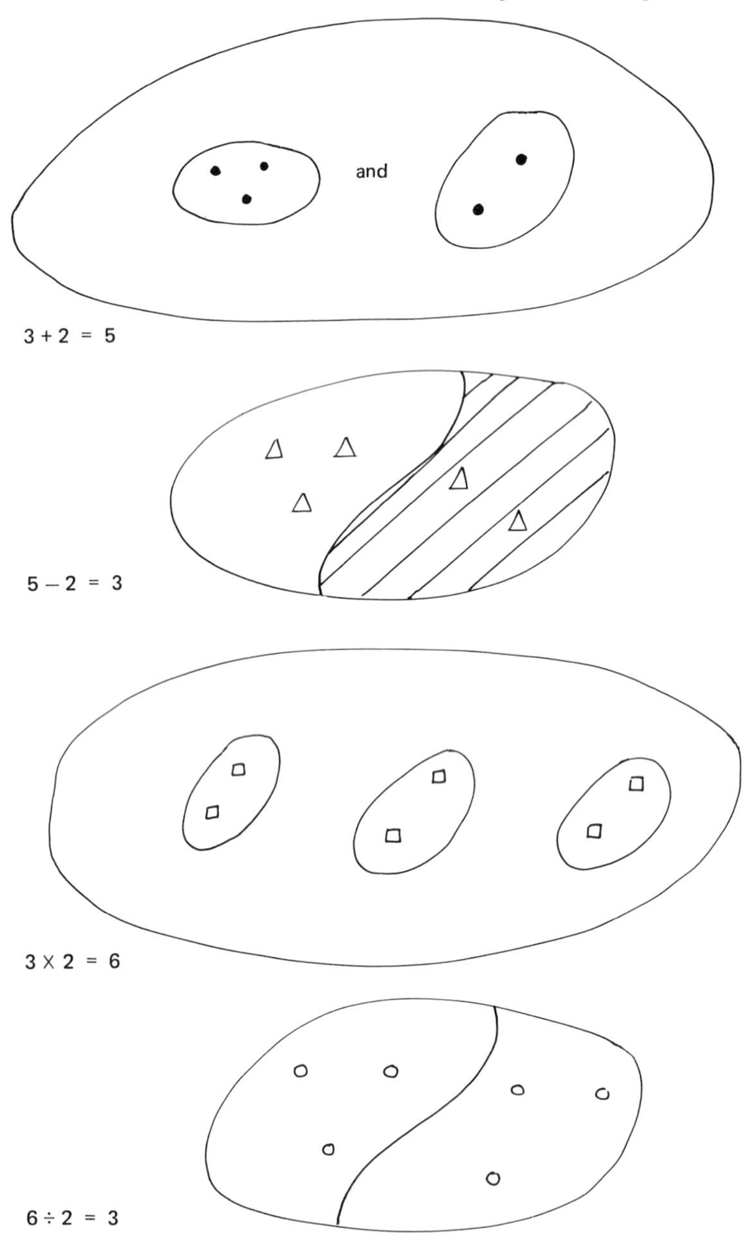

The number line is a natural development and it can later be extended to include negative numbers.

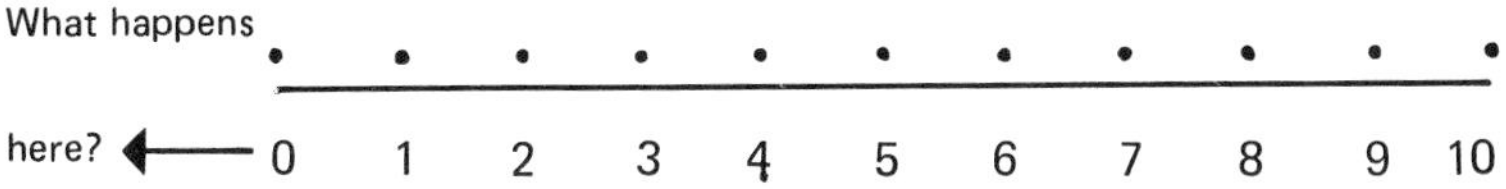

What happens here? ← 0 1 2 3 4 5 6 7 8 9 10

It is a good idea to cut out a strip of paper or card, labelling it as above, and to pin it to the wall of the classroom so that children can see and use it as they need but can also extend it in either direction as the need arises.

The understanding of 'place value' is one of the biggest stumbling blocks for slow learners. A lack of understanding of this concept at the junior stage leads to many misunderstandings at later stages. Secondary teachers of slow learners often find that they have to teach this again and it is then worth introducing concrete materials such as Dienes blocks in order to emphasise the concept. To acquire a real understanding of this concept it is essential that children should manipulate materials for themselves, and there is a natural progression when teaching this (see pp. 118-19).

The three approaches should now be used to teach the four rules of arithmetic for numbers greater than 10. It then becomes a natural procedure to write sums in symbolic form and children will understand how to do a sum such as 23 + 256 + 7 knowing that it can be written as

$$\begin{array}{r} 23 \\ 256 \\ +\quad 7 \\ \hline \\ \hline \end{array}$$

(1) *Sticks and bundles*

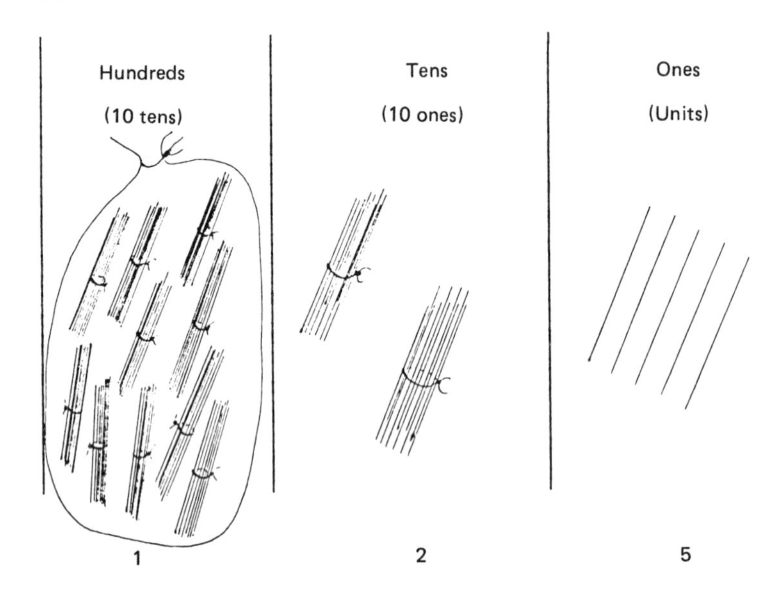

(2) *Dienes blocks*

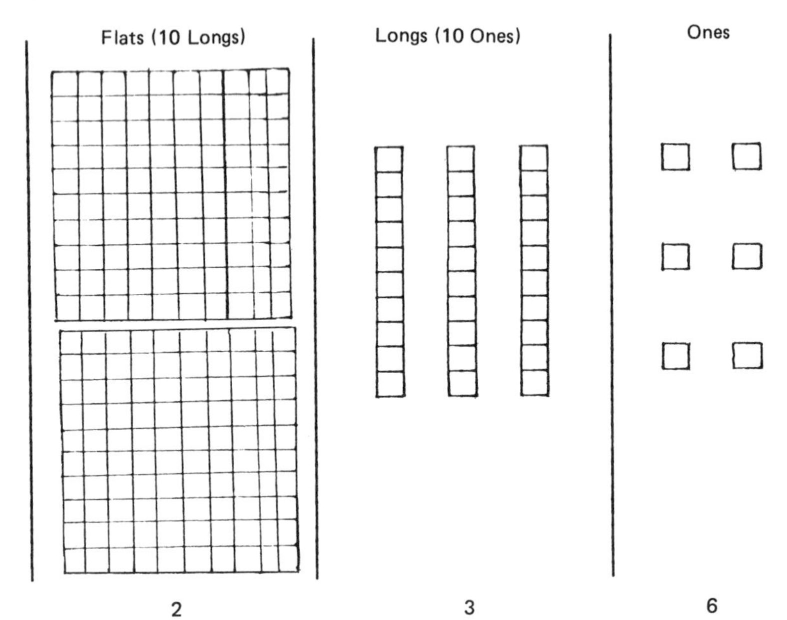

(3) *The Abacus*

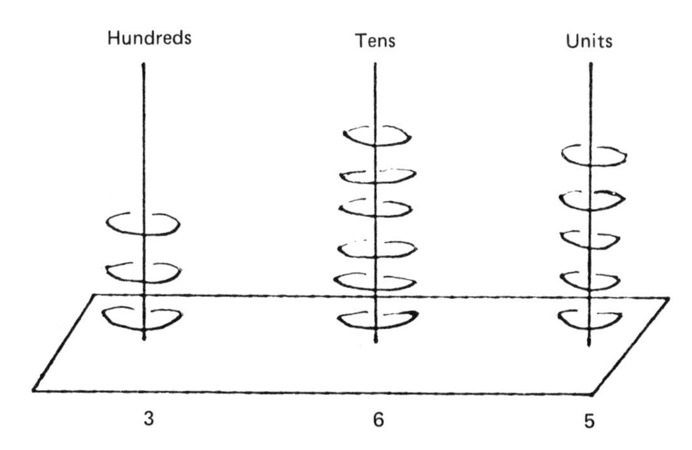

Furthermore the abacus can be used to illustrate decimals and sums involving the decimal point. Hopefully then, 39.6 + 2.75 + 4.538 will not be written as

$$\begin{array}{r} 3\,9.6 \\ 2.7\,5 \\ +\,4.5\,3\,8 \\ \hline \\ \hline \end{array}$$

Zero is always a problem and it is not really surprising to find a child who is asked to write 'forty three', writing '403', or when asked to write 'six thousand and fifty seven' he writes '600057', or if he is asked what '3005' stands for he says 'three hundred and five' (Hart, 1981). Frequent mistakes, too, are $3 \times 0 = 3$ and $3 \div 0 = 3$. These mistakes indicate a lack of understanding of the concepts of place value and zero. If a child is having such problems the use of Dienes blocks should help him to overcome them.

Fractions are too often taught with the assumption that they are merely an extension of the natural number system. They constitute an entirely different system. Natural numbers extend along a number line from zero to infinity. If a number is multiplied by another number the answer is *bigger* than the original and if a number is divided by another number the answer is *smaller*.

e.g.

$$5 \times 3 = 15 \text{ and } 15 > 5$$
$$24 \div 6 = 4 \text{ and } 4 < 24$$

Proper fractions only exist between zero and one,

$$0 \quad \tfrac{1}{8} \qquad \tfrac{1}{3} \qquad \tfrac{1}{2} \qquad \tfrac{3}{4} \qquad 1$$

but there is an infinite number of them. If a fraction is multiplied by another fraction the answer is *smaller* than the original and if a fraction is divided by another fraction the answer is *bigger*.

e.g.

$$\tfrac{4}{5} \times \tfrac{1}{2} = \tfrac{2}{5} \text{ and } \tfrac{2}{5} < \tfrac{4}{5}$$
$$\tfrac{3}{4} \div \tfrac{7}{8} = \tfrac{6}{7} \text{ and } \tfrac{6}{7} > \tfrac{3}{4}$$

In order that children should fully understand the concepts they should be given the opportunity to cut up real things (e.g. cakes or apples) and then pieces of card in various shapes – circles, squares, strips etc.

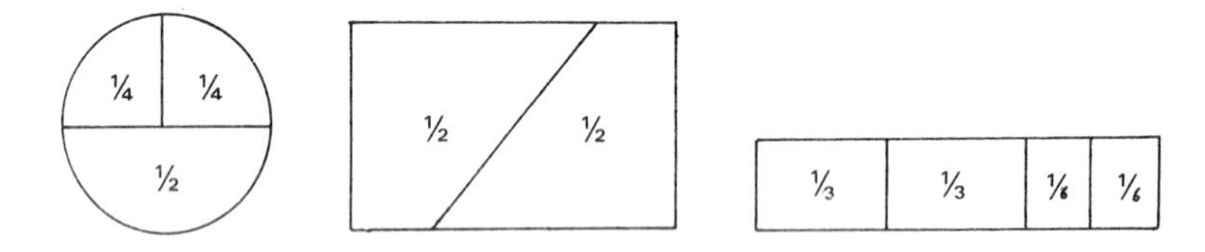

Much practice in cutting up shapes and fitting them together again is required. When teaching these ideas to slow learners it is only necessary to teach about halves and quarters (possibly eighths), thirds and sixths and possibly fifths and tenths. It is unnecessary to teach more complicated fractions and even these simpler ones should be seen to be relevant to, say, the sharing of an apple or some sweets, or buying a half-kilo of potatoes or a quarter of sweets, or cutting a length of wood into equal parts. The study of number is not just 'doing sums' and a lot of fascinating work can be developed by looking at number patterns.

The advantage of a study of this nature is that it can be viewed from a simple and basic point of view suitable for the slow learner or from a more sophisticated point of view suitable for the able child. Simple examples which could help a slow learner to appreciate and to 'become friendly' with numbers are given here:

(1) *Multiplication Tables*

Nine	*Tens column*	*Units column*	*Sum of digits*
1 × 9 = 9	0	9	9
2 × 9 = 18	1	8	1 + 8 = 9
3 × 9 = 27	2	7	2 + 7 = 9
4 × 9 = 36	3	6	3 + 6 = 9
5 × 9 = 45	4	5	4 + 5 = 9
6 × 9 = 54	5	4	5 + 4 = 9
↓ ↓	↓	↓	↓

Eleven	*Tens column*	*Units column*
1 × 11 = 11	1	1
2 × 11 = 22	2	2
3 × 11 = 33	3	3
4 × 11 = 44	4	4
5 × 11 = 55	5	5
↓ ↓	↓	↓

but

	Tens column	*Hundreds column + Units column*
11 × 11 = 121	②	1 + 1 = ②
12 × 11 = 132	③	1 + 2 = ③
13 × 11 = 143	④	1 + 3 = ④
↓ ↓	↓	↓ ↓

(2) *Square Numbers and Triangular Numbers*

A pegboard and pegs is a useful piece of equipment for this work.

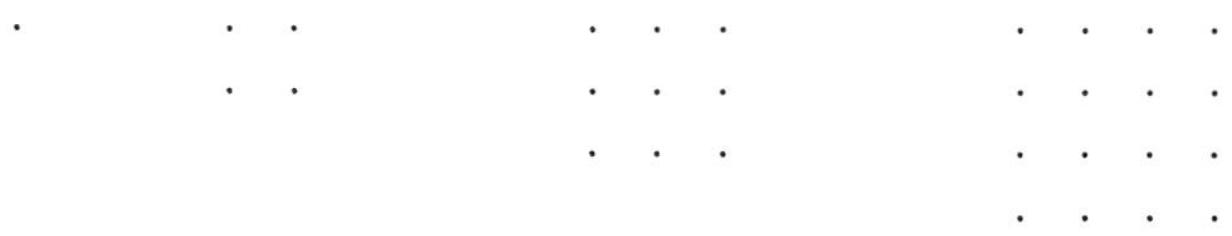

```
.        . .        . . .        . . . .
         . .        . . .        . . . .
                    . . .        . . . .
                                 . . . .
```

1	4	9	16
1	2^2	3^2	4^2
1st	2nd	3rd	4th

1	1 + 3	1 + 3 + 5	1 + 3 + 5 + 7

Such questions as 'What is the next number in the series?', 'What is the 7th number in the series?' should now be asked.

1st	2nd	3rd	4th
1	3	6	10
1	1 + 2	1 + 2 + 3	1 + 2 + 3 + 4

Similar discoveries may be encouraged with respect to these numbers. The study of pattern in number, whilst stimulating curiosity, encourages valuable learning.

Length and area are often not fully understood. In particular, the difference between the perimeter and area of a shape and the circumference and area of a circle. Children should have practice in walking round the edge of the classroom or school hall or playground. They should also be able to recognise that these distances are not the same as, for example, the number of tiles covering the hall floor or the number of slabs covering the playground. Too often area is taught in terms of a formula, the most obvious one being 'length times breadth'. Whilst this could be accepted as true in some cases it is certainly not what area really *is*. Irregular shapes (e.g. a leaf or a child's hand or foot) should be drawn on squared paper (having established that squares are the most sensible units to use) so that the concept of area may be understood.

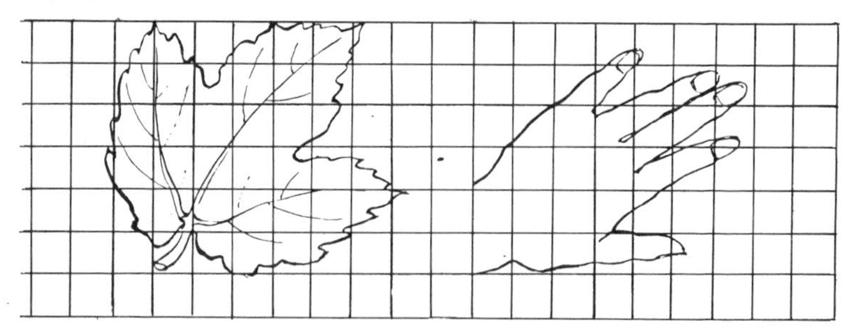

Children can then be led to discover some of the formulae for area. The geo-board is a useful piece of equipment for this purpose.

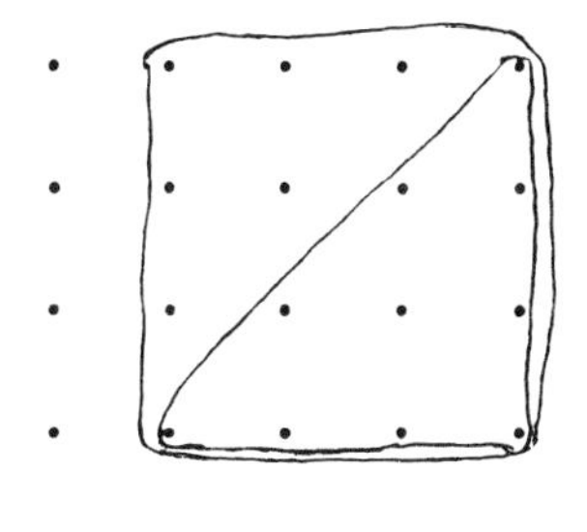

Area of the square is 9 squares
Area of the triangle is 3 whole squares + 3 half squares, i.e. 4½ squares
Area of the triangle is half the area of the square.

Geometry

There is no reason why slow learners should not learn some geometry and algebra as long as the introduction to them is performed in a concrete way. Glen (1979) suggests many excellent ways of introducing geometry using the environmental approach. The book was not written for slow learners but it gives some good and useful starting points for these children.

Comparison of size and shape is a basic idea which it is sometimes assumed need not be 'taught'. Slow learners do not 'pick up' ideas easily especially if they do not experience normal conversation at home. Hence, it *is* important that they are given opportunities to talk about the *tall* man and the *tallest* man and to compare the *big* car with the *bigger* car and the *biggest* car.

They should have plenty of opportunities to handle three-dimensional objects (a drawing of a cube on the blackboard may not look like a solid object to a slow learning child). Slow learners are quite capable of counting the number of faces, edges and vertices of various polyhedra, thus discovering Euler's equation, and of recognising the shapes of the faces.

Slow learners should also be encouraged to draw nets of solids and to construct their own polyhedra. This activity is a good one to do at the end of the Autumn term so that the polyhedra can be coloured by the children and then used as Christmas decorations. This sort of activity gives the children a feeling of success.

The idea of an 'angle' may be introduced by using a compass (*not* a pair of compasses) to find direction and noting that one *turns* from facing North to facing West through a right angle. This concept of an 'angle' as a 'turn' is important and it is then a simple step for the

Euler's Formula states that F + V = E + 2.

	Number of faces (F)	Number of edges (E)	Number of vertices (V)	Shape of faces	Value of F + V	E + 2
1 Cube (Vertex, Edge, Face)	6	12	8	Square	14	14
2 Tetrahedron						
3 Cuboid						

child to understand that the sum of the three angles of a triangle is 180 , or that the sum of the four angles of a quadrilateral is 360 .

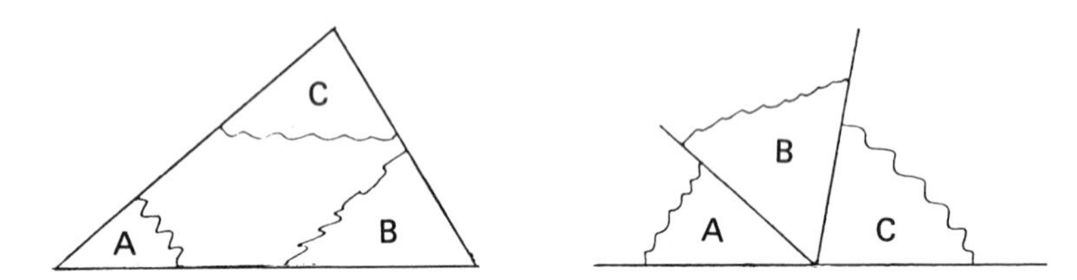

Diagrams, including graphs, form the stage in between the concrete and the formal stages. This is an important stage in a child's development and more use should be made of graphical work in the teaching of mathematics.

Children may be encouraged to collect their own data from their own environment. The usual example of 'counting cars' could be extended to a real environmental project – the local authority have to decide where to put traffic lights or whether to build a roundabout at the crossroads near the school. The children are asked to do a survey of traffic flow in, say, one hour:

(a) How many cars and lorries travel from East to West?
(b) How many cars and lorries travel from North to South?
(c) How many cars and lorries coming from the South turn West?
(d) Why do they get these results? E.g. a large city lies to the South and there is a sea port to the West.
(e) Draw graphs to illustrate these results.
(f) Draw diagrams to illustrate the flow of traffic.
(g) Make recommendations to the authority.

Other environmental projects could be devised so that children feel very much involved in important community decisions. It is a good thing if children can see how mathematics is not an 'isolated' subject but that it has a role to play in other subjects and in everyday life. Young people could be encouraged to think about the use of their time and money; graphs and diagrams can be used to illustrate these aspects of their lives.

Example: Draw a pie chart to show the use of pocket money.

Item	*Percentage of money*	*Angle at centre of circle*
Clothes	40	144°
Entertainment	30	108°
Fares	20	72°
Save	10	36°
Total	100%	360°

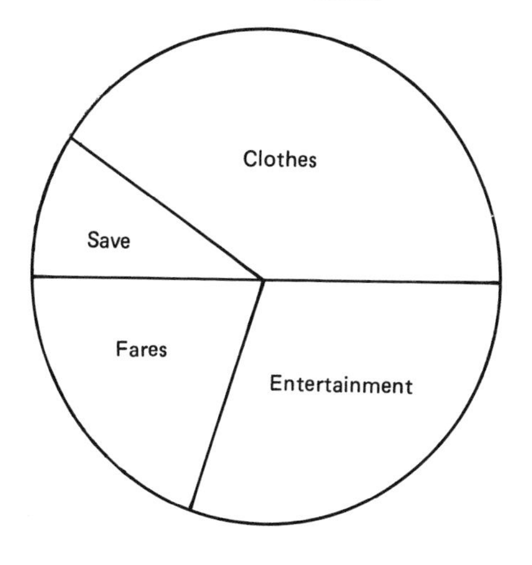

This exercise can involve a number of different aspects of mathematics – money, percentages, proportion, angles, areas of sectors, use of compasses and protractor and, finally, a value judgement is demanded when children are asked to discuss how they use their money.

Slow learners need to feel that they can be successful if they try and that someone cares about their success. They also need to feel that the work they do in school is worthwhile to them and is considered to be worthwhile by their teachers, their parents and the community. That is why mathematics should be seen to be useful, interesting and understandable. Until this is so we shall continue to see youngsters leaving school with adverse attitudes towards school generally and towards mathematics in particular.

6.5 The Able Child

The intense curiosity, imagination and creative ability of the very able child are the characteristics which should be encouraged and, indeed, exploited. It is important that a mathematically gifted child should be identified as early as possible, whilst, at the same time, he must be treated like a normal child. Intelligence tests are not the best way of doing this, nor is an objective measure of attainment, although these will obviously give some indication. Imagination, originality, creativity are different qualities from intelligence as measured by an IQ test which tends to give a final result without taking into account the process of attaining that result. Furthermore, an IQ test examines convergent thinking and since this sort of thinking is considered by many people to be respectable we have a plethora of intelligence and attainment tests. Divergent thinking, though, is related to creativity whilst not actually being equivalent to it. Whilst a high general intelligence (with a high IQ score) is usually part of giftedness, originality of thought, imagination and inventiveness and a desire to succeed are all important aspects of the character of the able child which will influence his learning and understanding of mathematics. Tests of creativity are available and should, if testing is required, be used alongside intelligence tests. They vary in kind but basically the child is asked to make up his own problems rather than being given a set problem to solve in a specific way.

It has been suggested (Rogers, 1918) that there are two aspects of mathematical ability – the *reproductive* relying upon memory and the *productive* which relies upon thought and imagination. Krutetskii (Krutetskii, 1976) distinguishes between *school* ability, involving the mastering and reproduction of mathematical information, and *creative* ability involving the creation of something original. As a result of his research he maintained that mathematical giftedness can show itself at a very early age when a child is seen to be able to manipulate numbers and is keen to do so. Particular characteristics of the mathematically able child are ability to generalise, to switch rapidly from one operation to another, to remember, to find the quickest and easiest way of solving a problem and a 'never tiring' attitude towards mathematical activities. It is also interesting that Krutetskii noted that, whilst some able children were able to solve problems in a logical way, others still felt happier with a visual means of interpretation. Some of Krutetskii's sample of children were generally gifted as well as mathematically able whilst others were quite ordinary children generally but showed a particular ability in mathematics.

It is important that these particular aspects and characteristics of mathematically able children are taken into account when a curriculum is planned for them. Forcing them to conform to the curriculum which is considered to be acceptable for the normal school population may cause much frustration and this can lead to a rejection of the school's norms of behaviour and even to rejection of society's norms of behaviour.

The Mathematically Gifted Child in the Primary School

Primary education has changed over the past 20 to 30 years. The classroom is now a relaxed and stimulating place. All children need to have freedom to move around and to discover for themselves and, in particular, 'able children thrive in the open country of learning' (Hopkinson, 1978). Inevitably the number of able children in any one class is likely to be small (top 10 per cent perhaps?); sometimes the mathematical background of the primary school teacher is scanty and he/she lacks confidence and imagination. This often leads to an excessive use of text books and the stifling of the able child's imagination. This should be avoided at all costs.

Children of all abilities tend to be interested in practical work and the teacher should avoid falling into the trap of thinking that able children 'do not need concrete materials – they prefer to think in the abstract'. It is essential that the teacher should be able to ask questions about the work the children are engaged in and to encourage able children to find out and to develop new ideas. Examples of this practical work are:

(1) The use of the equilibrium balance and the balance board leading to concepts of balance, of moments and of equations

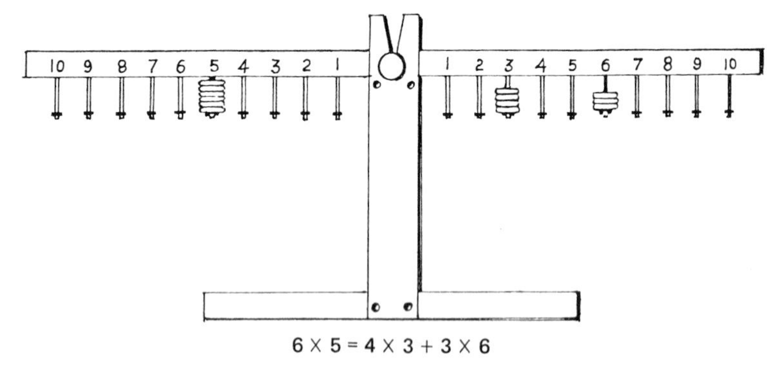

$6 \times 5 = 4 \times 3 + 3 \times 6$

(2) The study of the concept of time, the use of the pendulum and the water clock and the construction of 'time-pieces'. The tabulation of results also becomes important.

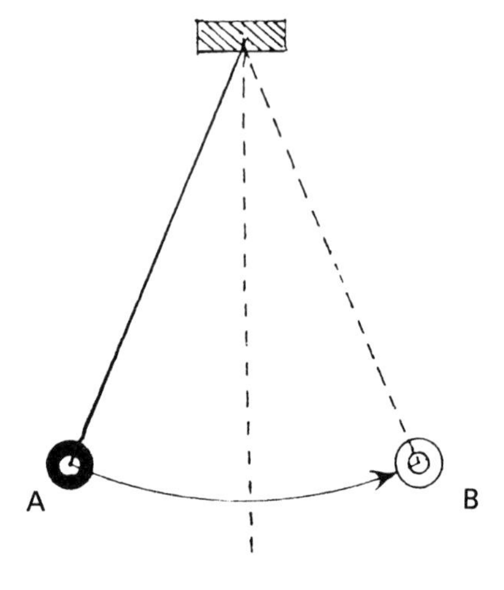

The time taken to swing from A to B depends upon the length of the string and not upon the weight attached to it.

(3) The concept of area using tangrams and other Chinese puzzles. The study of tesselations can be a fascinating exercise when taken beyond the simple tesselating shapes, e.g. the art of Escher.

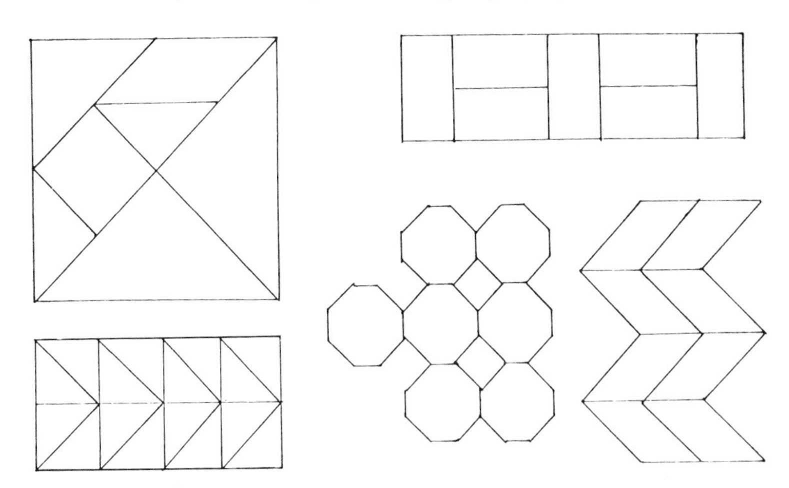

Mathematical thinking needs to be stimulated in the able child and the following examples help to do this. Furthermore, they may be seen as further developments into abstract thought, of work already being done by the whole class:

(1) *Number Patterns*

·	· ·	· · ·	
	· ·	· · ·	-------------→
		· · ·	
1	4	9	------------- nth square?
1	2^2	3^2	------------- n^2
1	1 + 3	1 + 3 + 5	------------- 1 + 3 + 5 + ---- (2n − 1)
1st	2nd	3rd	---------------- nth

(2) *Logic*

The use of Dienes' Logi-blocs is important for all children. Very able children will often work with these materials making up their own quite complicated rules. Soon they can be given examples to do which do not involve the use of the materials.

(3) *Spatial Concepts*

The construction of three-dimensional models is an activity which often fascinates the more able child. There are some interesting publications giving instructions on how to make geometrical models and included in these books one can find the geometrical facts and proofs which emerge (see Cundy and Rollett, 1961 and Wenninger, 1971). Primary children may be led to their own limit of understanding and this work may well be continued when they are older.

Access to interesting books to encourage children to do further research is important. Books like *Man Must Measure* (Hogben, 1937), *The Language of Mathematics* (Land, 1960) and *Children Learning Geometry* (Glenn, 1979) are all valuable for this purpose. It is also a good idea to make available books showing the historical development of mathematics. Suitable ones for children at primary level are *Number Stories of Long Ago* (Smith, 1919) and, possibly, *A History of Mathematics* (Freebury, 1958).

Able children must not always be distinguished as 'different' and so it is important that they should often work within a group of other children, perhaps when they are carrying out a particular project. Invariably they prove to be valuable members of a group and one can see their influence working with other children who then think and learn more than they would otherwise have done.

Examples of this sort of work are:

(1) the use of measuring instruments, trundle wheels, maps etc. to measure and then to draw plans
(2) the construction of polyhedra.

A variety of mathematical games should be available for children to enjoy and, at the same time, to exercise their intellect. Children will often make up their own rules.

The Mathematically Gifted Child in the Secondary School

There are various ways of arranging the teaching of a mathematically gifted child. He may be placed in a special class in his normal school; he may be subjected to a form of *acceleration* i.e. he is placed in a class of pupils who are older than he is; a special *enrichment* programme may be planned for him. If his particular ability is only in mathematics, then acceleration is not suitable as he may find himself 'out of his depth' in other subjects. Furthermore, the development of his whole personality must be considered; it is not necessarily satisfactory for a child of, say, eleven years to be educated with children of fourteen years, for physically and emotionally an eleven-year-old is different from a fourteen-year-old. Placing children in special classes immediately segregates them and labels them as 'different'. This is a pity because able mathematics pupils not only have a lot to contribute to normal mathematics lessons but also they can still learn a great deal from mixing with average pupils. The occasional *extra* or *special* lesson given by an able teacher of mathematics may well be beneficial, though. Able children need to come into contact with good mathematical brains.

A special enrichment programme would usually seem to be the answer. The child will follow the normal mathematics curriculum but should be given extra 'experiences'. A teacher must be sensitive to the child's requirements and be prepared to search for new materials and ideas. A good library and the use of some television and radio programmes are essential, and children should be directed to appropriate books and programmes. The calculator is also a useful piece of equipment as the 'drudgery' of arithmetic can then be avoided. Sometimes a computer will arouse interest. The rapid development of micro-computers implies that most secondary schools will soon have at least one, and able children, as well as less able, may be stimulated by their use. However, it is said that the computer relies upon convergent thinking and could, therefore, be a frustrating experience for the divergent thinker. Many able children are interested in 'how things work' and so the relationship of mathematics with the sciences should also be stressed.

To encourage the aesthetic side of a child's personality there are many interesting books relating mathematics to art, music, archaeology etc. and a child will often become deeply involved with such subjects.

What Mathematics?

A burning question seems to be 'Do we teach the mathematically gifted child *more* of the same mathematics that we teach to other children or do we teach them something different?' This is not really an 'either/or' question for both aspects should apply. Whilst the class is studying Pythagoras' theorem the able child might be encouraged to (1) find out about the life of Pythagoras, his secret society of Pythagoreans and some of the work they did; (2) to devise other proofs of the theorem; (3) to discuss applications of the theorem and (4) to discuss the nature of 'proof'. (See Baxandall, Brown, St. C. Rose and Watson, 1978.) These four channels may lead to many other studies. One can see that they all relate to the work being done by the rest of the class but, at the same time, they are encouraging the able child to make further discoveries for himself.

When the class is studying numbers and number patterns the able child could be encouraged to find proofs for general results possibly using mathematical induction or look at the way number concepts have developed historically (Smeltzer, 1953). Some work with complex numbers may also be of interest.

Graphical work can lead a child along many roads including the study of the concept of infinity and infinite series, and of limits and convergence. Sawyer (1970) tackles some of these problems in practical ways that would excite any budding mathematician and Land (1960) sees mathematics from the viewpoint of its applications to ordinary, everyday phenomena.

For the child who is interested in art as well as mathematics a study of pattern in nature, perspective, the golden section and the Fibonacci series, architecture and some of Escher's art could all help him to develop his enquiring mind and aesthetic appreciation both at the same time.

As Budden (1981) points out it is important that we make the distinction between gifts, talents and skills:

> Gifted normally means endowed naturally with intelligence or other inborn potential; talented implies, I feel, that such gifts have been exercised and developed; skilled usually refers to a motor skill.

If we wish to produce 'talented' mathematicians we must first recognise the 'gifted' one and then exercise and develop such gifts as he possesses.

6.6 Conclusion

> It is probably true to say that mathematics is one of the disciplines where there is the largest range of ability. Second only, perhaps, to music, where the range is from Mozart down to the tone-deaf – a truly astronomical ratio. I would claim that in mathematics, the gap also deserves to be called astronomical. (Budden, 1981)

Discussion, in this chapter, has centred on the two extremes of the ability spectrum of children in normal schools. Some suggestions of topics to be taught have been made, but the teacher will need to use his/her own initiative to develop other topics and ideas.

Both the slow learners and the gifted children need careful, skilled and sympathetic teaching in mathematics if their potentials are to be realised. The brighter children *cannot* look after themselves and a lot of harm *can* be done to slow learners by bad or unimaginative teaching. Inspired teaching can produce in children of all abilities a love and enthusiasm for mathematics and this should be the aim of all teachers of mathematics, whether they are teaching gifted children, 'normal' children or slow learners.

Bibliography

Baxandall, P.R., Brown, W.S., St. C. Rose, G. and Watson, F.R. (1978) *Proof in Mathematics*, University of Keele.
Berrill, R.A. (1980) 'Attitudes Towards Mathematics of Pupils and Teachers in Secondary Schools', unpublished PhD thesis, Newcastle University.
Biggs, J.B. (1967) *Mathematics and the Conditions of Learning*, NFER, Slough.
Budden, F.J. (1981) 'Some Thoughts on Gifted Children', *Mathematics in School*, vol. 10, no. 1.
Cundy, H.M. and Rollett, A.P. (1961) *Mathematical Models*, Clarendon, Oxford.
DES (1978) *Special Educational Needs: Warnock Report*, HMSO, London.
DES (1979) *Aspects of Secondary Education in England*, HMSO, London.
DES (1980) *Aspects of Secondary Education in England: Supplementary Information on Mathematics*, HMSO, London.
Dienes, Z.P. (1960) *Building up Mathematics*, Hutchinson, London.
Dienes, Z.P. (1973) *The Six Stages in the Process of Learning Mathematics*, NFER, Slough.
Donaldson, M. (1978) *Children's Minds*, Fontana, London.
Freebury, H.A. (1958) *A History of Mathematics*, Cassell, London.
Glenn, J.A. (1979) *Children Learning Geometry*, Harper and Row, London and New York.
Hart, K.M. (1981) *Children's Understanding of Mathematics: 11-16*, John Murray, London.
Hogben, L. (1937) *Man Must Measure*, Allen and Unwin, London.
Hopkinson, D. (1978) *The Education of Gifted Children*, Woburn Educational Series, London.
Inhelder, B. and Piaget, J. (1966) *The Growth of Logical Thinking from Childhood to Adolescence*, Routledge and Keegan Paul, London.
Krutetskii, V.A. (1976) *The Psychology of Mathematical Abilities in School Children*, University of Chicago Press, Chicago and London.
Land, F. (1960) *The Language of Mathematics*, John Murray, London.
Lovell, K. (1971) *The Growth of Understanding in Mathematics Kindergarten through Grade Three*, Holt, Rinehart and Winston, New York.
National Council of Teachers of Mathematics (1972) *The Slow Learner in Mathematics*, Year Book No. 35, NCTM, USA.
Pringle, M.K. (1975) *The Needs of Children*, Hutchinson, London.
Rogers, A.L. (1918) *Experimental Tests of Mathematical Ability and their Prognostic Value*, Teachers' College Contribution to Education, New York.
Sawyer, W.W. (1954) *Mathematician's Delight*, Pelican, London.

Sawyer, W.W. (1970) *The Search for Pattern*, Pelican, London.
Skemp, R.R. (1971) *The Psychology of Learning Mathematics*, Penguin, London.
Smeltzer, D. (1953) *Man and Number*, Black, London.
Smith, D.E. (1919) *Number Stories of Long Ago*, Ginn, Boston, Mass.
Wenninger, M.J. (1971) *Polyhedron Models*, Cambridge University Press, Cambridge and New York.
Williams, A.A. (1970) *Basic Subjects for the Slow Learner*, Methuen, London and New York.

7 CALCULATORS IN THE CLASSROOM

Colin Noble-Nesbitt

> 'You are undoubtedly the best mathematics teacher in your school but wouldn't the occasional use of calculators make your lessons even more interesting and effective?'
>
> (Imaginary comment from a Maths teacher)

7.1 The Effect of External Examinations

Secondary Schools

Pebbles, abaci, Napier's 'bones', logarithms, slide rules and mechanical calculating machines, plus an array of pencil-and-paper algorithms: formidable testimony to mankind's inventiveness in gaining increasing mastery over arithmetical calculations. Add the electronic calculator to that list and reflect upon the gigantic step forward. And yet history may well look back to the 1970s as a decade when the vast majority of mathematics teachers in our secondary schools failed to respond effectively to the possibilities opened up by this new technology. This failure may be partly due to the ascendency of logarithms, Napier's invention, as the main aid to calculation in most external examinations at 16+. Perhaps an educational system like our own, but operating during Napier's lifetime, would have fossilised his 'bones' into its mathematical syllabuses and rejected his more potent invention! In direct contrast, the calculator has been welcomed with open arms by industry, commerce and the public at large. It is therefore reasonable to hope that this assimilation can be helped by what takes place in our schools during the 1980s.

As we turned into the 1980s, encouraging signs could be noticed in that an increasing number of GCE boards were moving, or had already moved, towards allowing the use of calculators in their mathematics papers at Ordinary Level. However, optimism should be tempered by the fact that, with the notable exception of one syllabus from the School Mathematics Project, the calculator was viewed as an alternative, but not essential, aid to calculation; syllabus design had not changed to match the exciting new power available at anyone's fingertips for less than the price of a pair of football boots. Indeed, it became a matter of amusement to some teachers that claims were being made that mathematics papers could be designed to confer no significant advantage to candidates who elected to use calculators! If this were the case in examinations designed for the most mathematically able quarter of our school population, then surely the picture could be expected to be somewhat better for those pupils who were entered for CSE mathematics examinations. After all, these pupils generally have a lesser facility for number crunching and the calculator becomes something of a godsend for them. But no! A survey of 14 CSE boards, in 1979, showed that only two of them allowed any use of calculators in their mathematics examinations. Since the mathematics teachers in our secondary schools control CSE mathematics syllabuses, they must accept much responsibility for this particular failure to respond more positively to the needs of the majority of their pupils in relation to the advent of calculators. One of the reasons put forward for this slower acceptance of calculators in CSE examinations as compared with Ordinary Level examinations concerns calculator availability. Thus an eight form entry comprehensive school might typically enter 140 pupils for a CSE examination and 60 pupils for a GCE examination; for the former the problem of adequate calculator supply by the school is generally considered to be insurmountable in terms of capital cost, whereas for the latter it is often expected that pupils should provide their own, with a small emergency back-up by the school. This is tantamount to treating CSE mathematics pupils as of second-class status – are there really such insurmountable obstacles in the way of expecting these pupils also to supply their own calculators? Indeed, most CSE pupils already own a calculator, or at least have access to one at home, and often use it to check or do homework, no matter what their teachers might hope. Calculators are with us and will not go away!

Mismatches between everyday reality and a cloistered mathematics curriculum from about 14+, fed by external examinations in which the calculator is ignored altogether or relegated to a role on a par with the use of tables, seem to have a far reaching stifling effect upon calculator use as an aid to teaching during the early years in secondary schools. Acceptance of the calculator as an essential element in external examinations at 16+ is a national necessity, both in itself and as a catalyst for CALIM (Calculator Assisted Learning in Mathematics). With a single system of examining mathematics at 16+ in prospect towards the end of the 1980s, it is to be hoped that the national criteria for such a system will encourage the development of CALIM and accelerate its spread to all secondary schools.

Because calculators are generally accepted in mathematics papers at GCE Advanced Level, they have become the means of carrying out nearly all but the most trivial calculations for the vast majority of pupils in sixth forms. The effects of this change are perceived most easily in statistics – calculators give this branch of mathematics a tremendous boost by virtue of speeding one towards a result and leaving more time for reflection upon the associated statistical principles and inferences. There are also encouraging signs that the calculator is having some impact as a teaching aid during the introduction and consolidation of topics in pure mathematics at this level.

Primary and Middle Schools

The removal of 11+ examinations from most parts of the country has given primary schools a freedom, in theory at least, to make changes in their approaches to mathematics in general and basic numeracy in particular. Unfortunately, in many primary schools, basic numeracy has long been equated with skill in carrying out the processes of written computation, as exemplified by the long multiplication and long division algorithms. This type of popular view of basic numeracy is shared by the majority of the public at large – quite naturally, because they have passed this way themselves!

In order to make progress, primary school teachers may find it useful to stand back and view pencil-and-paper algorithms in historical perspective. Most teachers will know from the mathematical component of their professional training that the system of Roman numeration is well suited to the recording of calculations performed on an abacus but not at all suited to the performance of the long multiplication and long division pencil-and-paper algorithms. The latter depend upon the decimal system of numeration for their efficient operation and owed their gradual dominance in the Western world to the need of trades-people for a more portable and adaptable means of calculation than was afforded by the abacus. It should be remembered of course that many abaci provide a concrete way of representing numbers in the decimal notation and they laid the foundation for the ingenious search, led by Pascal and Leibniz, for the best mechanical driven abacus. The mechanical desk calculator was the final embodiment of this search and subsequent development. It is such machines, along with slide rules, logarithms and pencil-and-paper algorithms, which have so recently been made redundant by electronic calculators (and computers), almost everywhere but in our schools.

Although it has been possible to retain the excellent decimal system as the means of input and output for the user, the internal operation of an electronic calculator is dependent upon an adaptation of the binary system of numeration, known as *binary coded decimal*, in which each separate decimal digit is represented by four binary digits e.g. the number 69 is represented as 0110/1001. The modern calculator can thus be thought of as an electronically driven abacus which operates in base two. This profound simplification has given us a means of computation which is more than a match for any which have gone before. The 1980s is surely the time for all those who have responsibility for, or influence on, the teaching of mathematics in our primary schools, to make more than a passive acknowledgment of this immensely powerful means of processing numerical data. Primary school teachers may not yet be ready to accept 'the intelligent use of a calculator' as a reasonable definition of basic numeracy but they might at least be expected to direct part of their mathematics teaching time to ensuring *some* intelligent use of a calculator for *all* of their pupils.

Middle schools, typically taking pupils in the age range 9 to 13, share with primary schools their general freedom from external examinations while often enjoying the additional advantage of having specialist mathematics teachers. This makes such schools a fertile ground for the introduction and imaginative development of CALIM.

7.2 Producing Classroom Materials

In order to make real headway with CALIM, many teachers will feel a need for specific pupil material, accompanied by teachers' notes and capable of fairly direct implementation in the classroom. In the longer term, new text-book based courses may emerge to meet this need to the full. Meanwhile, some teachers may like to try producing their own material. For the purposes of showing possible approaches to this task, two examples follow; in the first example, 'Pythagoras' Theorem and its Proof', pupil material and teachers' notes are given, whereas in the second example, 'Best Buys and Unit Pricing', the pupil material is only partly developed (visit a supermarket to provide further, up to the minute, examples) and the teachers' notes are left for the teacher to provide. The first example indicates one way in which a calculator can open up a new avenue to a very familiar piece of school mathematics, using a guided discovery approach; the second example draws attention to 'consumer applications' as a fertile source of real-life situations which can at last be made meaningful to most children.

Pythagoras' Theorem and its Proof: Pupil Material

Stage 1. Copy Table 7.1

Table 7.1

(1)	(2)	(3)	(4)	(5)	(6)
Areas of the two smaller squares		Areas of the two rectangles		Column (1) plus Column (2)	Area of largest square
P (cm^2)	Q (cm^2)	R (cm^2)	S (cm^2)	P+Q (cm^2)	R+S (cm^2)

Stage 2. A right angled triangle is shown in Figure 7.1. Squares have been drawn on each of the sides. The broken line divides the largest square into two rectangles, R and S.
Lengths are given in centimetres (measured to the nearest millimetre).

Stage 3. Calculate the areas of square P and rectangle R.
Write the results in columns (1) and (3) of your table.
How close to one another are these results?
Could P and R possibly have equal areas?
Your teacher will talk about this.

Stage 4. Do the same for the areas of square Q and rectangle S.
Write the results in columns (2) and (4) of your table.
Could Q and S possibly have equal areas?

Stage 5. Complete columns (5) and (6) of your table.

Stage 6. Make measurements on the new right-angled triangle in Figure 7.2 and complete the second line of results in your table.
Can P and R possibly have equal areas this time?
What about Q and S?

Figure 7.1

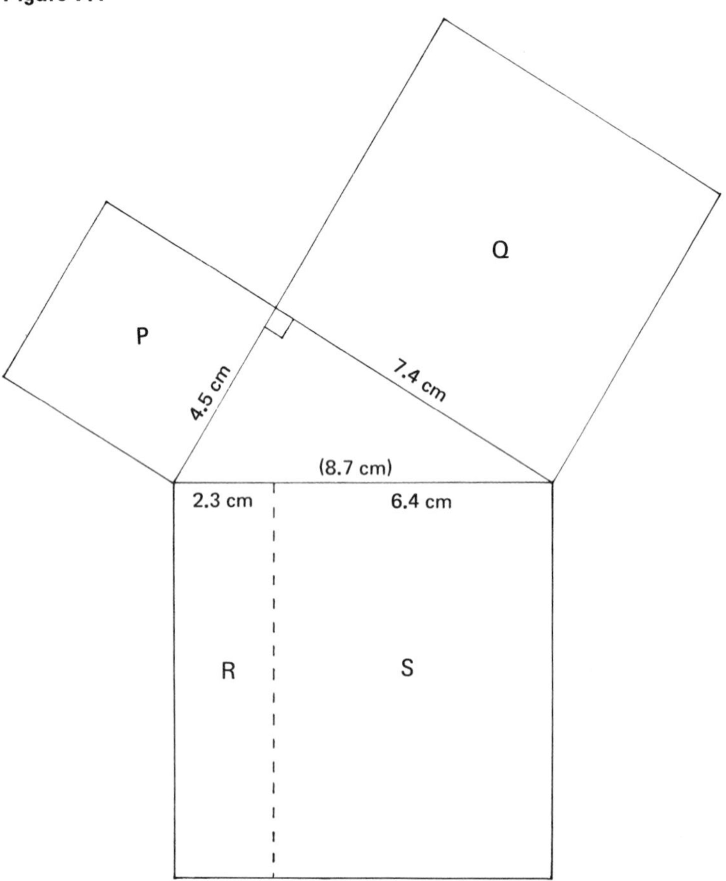

Stage 7. Using squared paper, draw a right-angled triangle of your own – the larger the better.
Make measurements so that you can complete the third line of your table.
Can P and R possibly have equal areas this time?
What about Q and S?

Stage 8. Compare column (5) with column (6). Is there anything which you notice to be approximately true?

Figure 7.2

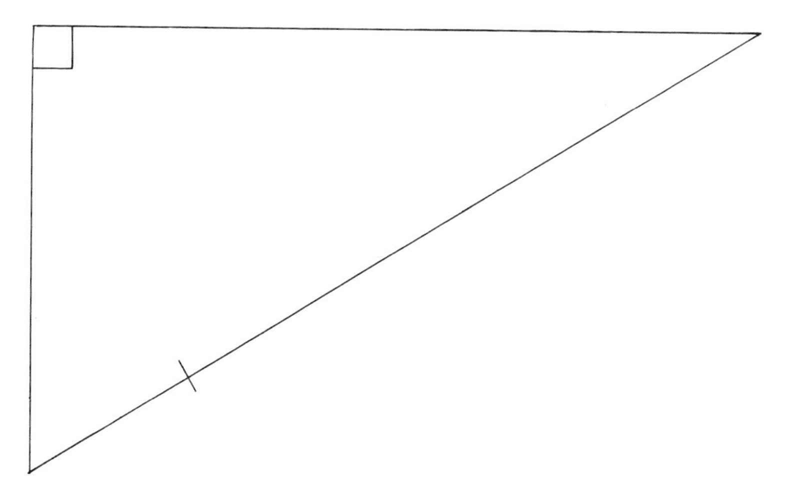

Pythagoras' Theorem and its Proof: Teachers' Notes

Purposes. (1) To indicate the geometrical relationship known as Pythagoras' rule. (2) To motivate a particular proof of the rule.

Target Groups. Upper ability second and third year secondary school pupils.

Materials Needed. Calculators, squared paper and set squares, metric rulers.

Prior Knowledge and Skills To Be Practised. Accurate measurement of length (to the nearest millimetre). The calculation of the area of rectangles (including squares). The analysis of errors in results calculated from physical measurements of length. The invariance of area under the plane transformations of shearing and rotation.

Preliminary Comment. Pythagoras' Theorem can be demonstrated or proved in an astonishing number of ways. In its application, it is most often used as a numerical statement about the *lengths of the sides* of a right-angled triangle. This numerical statement, corresponding to the formula $h = \sqrt{a^2 + b^2}$, can be easily verified by measuring the lengths of the two smaller sides of a pupil chosen right-angled triangle, calculating 'h' and finally checking the forecast length of the hypotenuse. This direct experimental approach, in which the calculator is a powerful influence, may be ideal in its simplicity for many pupils; it has the attraction of not complicating the final view by explicit consideration of the *areas of the squares* drawn on the sides of the triangle. The present pupil material differs in providing an experimental approach which motivates the need for a particular form of geometrical proof, summarised in the sequence of diagrams of Figure 7.3.

Figure 7.3

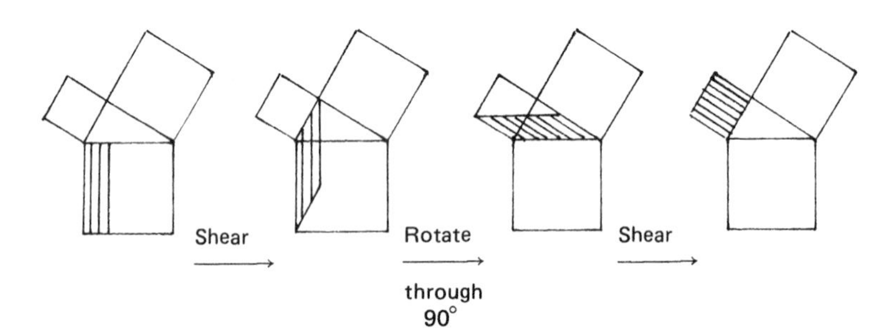

Read Before Stage 1. Duplication of the table for pupil use will save lesson time.

Read Before Stage 2. The broken line is perpendicular to the hypotenuse and passes through the vertex of the right-angle, when extended. The accuracy of the measured lengths is significant when discussing errors in the calculated areas of Stage 3. Thus a measurement of a = 4.5 cm, to the nearest millimetre, corresponds to the statements $4.45 \text{ cm} \leqslant a < 4.55 \text{ cm}$ and $19.802\,5 \text{ cm}^2 \leqslant a^2 < 20.702\,5 \text{ cm}^2$ (or $19.8 \text{ cm}^2 < P < 20.8 \text{ cm}^2$). In a similar way, $19.4 \text{ cm}^2 < R < 20.6 \text{ cm}^2$.

Read Before Stages 3 and 4. The calculated values of $P \simeq 20.3 \text{ cm}^2$ and $R \simeq 20.0 \text{ cm}^2$ can form the basis of a useful (or even essential) class discussion of the question – 'Could P and R possibly have equal areas?' This discussion can ultimately motivate the proof as summarised diagrammatically in Figure 7.3.

Read Before Stage 5. There does not seem to be any need to interrupt the flow of pupils' work at the end of this stage.

Read Before Stage 6. Now that the squares and rectangles have been seen in position in Figure 7.1, it is hoped that they can be left to the imagination in Figure 7.2, allowing for a larger right-angled triangle with consequent lower relative error in the measurements of length.

Read Before Stage 7. It is useful to provide some pupils with really large pieces of squared paper in order to convince them that their findings do apply to right-angled triangles of any size.

Read Before Stage 8. Comparison of column (5) with column (6) should provide the basis for the conjecture that the area of the square drawn on the hypotenuse of a right-angled triangle is equal to the sum of the areas of the two squares drawn on the other sides of the triangle. This conjecture calls for a proof; the structure of the pupil material leads naturally to the one in which a knowledge of shearing and rotation is used, as mentioned in the preliminary comment.

Best Buys and Unit Pricing: Pupil Material

Emma and Peter were shopping in the local supermarket. They needed a tin of peas. On the labels of two different sized tins, they found this information:

Green Stalk	Green Stalk
Garden Peas	Garden Peas
17p	28p
283 g	538 g
10 oz	1 lb 3 oz

Peter wanted to know how much they each cost for 1 Kg (= 1000 g). His calculator told him (see Figure 7.4):

Small tin	Large tin
60p per 1 Kg	52p per 1 Kg

Check Peter's calculations. Emma put the large tin in her basket. Did she choose the best buy?

Some people think that UNIT PRICES should be put on labels,

Figure 7.4

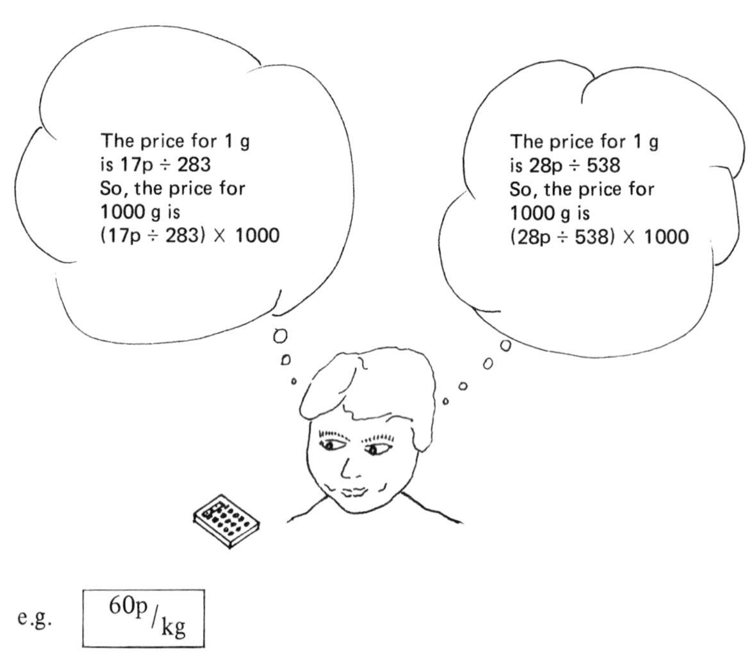

e.g. 60p/kg

on the small tin. Do you agree?

Further on, Emma noticed a different brand of peas, in tins of two different sizes. The labels gave this information:

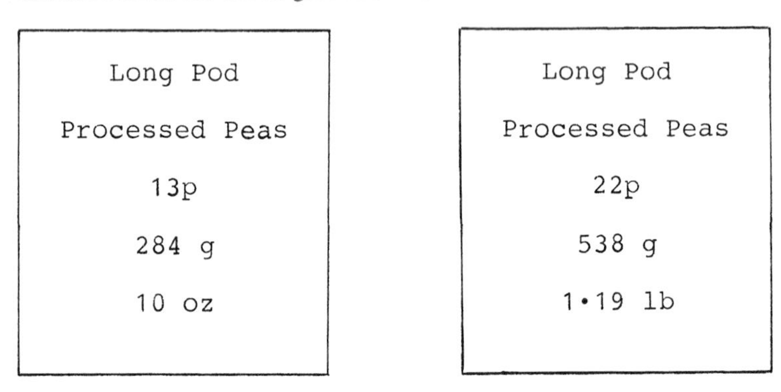

Work out the UNIT PRICES for the peas in these tins. Emma replaced the garden peas! Peter agreed that they should investigate further. They found a third brand of peas. This time the information was:

Green Wonder	Green Wonder
Processed Peas	Processed Peas
12p	18½p
298 g	539 g
10½ oz	1 lb 3 oz

Work out the UNIT PRICES for these tins. Which tin of peas would you choose to buy, and why?

7.3 Calculator Use from 11 to 16

Some reorganisation of the secondary mathematics curriculum is inevitable if the full potential of calculators is to be realised. Nevertheless it seems unlikely, even ten years on, that many of the topics which are in the present syllabuses will have been replaced by completely new topics. Perhaps from the viewpoint of pupil interest and relevance to today's computer world, iterative methods need consideration for inclusion, at least for more able pupils. The idea of an algorithm also seems ripe for development with the emphasis swinging away from some of the familiar pencil-and-paper procedures to those which are designed for use with a calculator. Although the need for 'the appropriate degree of accuracy' has long been called for when answering examination questions, the response by teachers has generally been to prime their pupils with expedient rules of thumb which could not possibly meet the requirements of every situation; the calculator makes it possible to plan a much more considered approach to this whole question of the relationship between the accuracy of a final result and the natural errors which are normally present in the original data.

With a gradual change of emphasis in prospect for mathematics syllabuses, rather than wholesale change, teachers can concentrate on how the calculator can support their teaching of many topics found in current syllabuses. The purpose of the following notes is to focus attention on some of these topics and to call attention to the impact calculators can have when dealing with them in the classroom. Where thought to be relevant, comment is included regarding particular features which are available on some calculators.

Finding Factors of Whole Numbers

For many pupils, a pencil-and-paper approach to this topic only serves to emphasise deficiencies in their recall of basic multiplication facts and the concept of a factor becomes obscured. The casual or systematic use of a calculator, on the other hand, can help such pupils to appreciate the concept and in an incidental way may even enhance their ability to recall multiplication facts. A systematic calculator search can also be used in finding prime factors and then the prime factorisation of a large range of numbers becomes possible in an enjoyable way.

Repeated Addition and Repeated Subtraction

Where the fundamental relationship between multiplication and repeated addition needs refreshing in the minds of some pupils, the calculator is a useful aid; its use is even more potent when looking at division as a short-cut to repeated subtraction. In these applications it is very helpful if the calculators used have what is known as a 'constant facility' on both addition and subtraction. This allows the user to key in the first addition (or subtraction) after which extra presses of the '=' key produce the results of subsequent additions (or subtractions) without having to again enter the number which is being repeatedly added (or subtracted).

Place Value in Decimals

The effects of multiplying or dividing a given number by 10, 100, 1000, . . . are clearly evident in a calculator display and this can help pupils to gain a clearer idea of the importance of the position of a digit in relation to the decimal point.

Comparison of Size of Vulgar Fractions

Conversion to decimal form, using a calculator, has become the quickest way of carrying out the comparison and has the bonus of providing experience of numbers with a lot of decimal digits (up to seven with an eight digit calculator display), emphasising the place-value concept to the right of the decimal point. In any given comparison of two fractions, the corresponding decimals only need to be compared up to a certain number of significant figures or decimal places, providing an opportunity for discussion of these two further concepts also. Here are two pairs of fractions to compare, using a calculator: $^{2}/_{7}$ and $^{104}/_{365}$ (the fraction of the year occupied by weekends?); $^{3927}/_{1250}$ and $^{377}/_{120}$ (two rational approximations to π). How many significant figures are needed to carry out the comparisons?

Pie Charts

Text-books have almost always introduced and developed the idea of a pie chart by using examples in which the circles were effectively divided into 'n' equal parts, with 'n' a factor of 360. If a class of pupils happened to have, for example, 29 or 31 members it was unlikely that their teacher would be attracted to the idea of asking them to construct a pie chart showing how they travelled to school! With calculators available the position is quite different – the computation becomes easy and the effort is transferred to interpreting the significance of the digits which appear in the calculator display, again developing pupil understanding of place value.

Directed Numbers

It is important to make sure that calculators chosen for classroom use have a means of entering negative numbers as well as a way of displaying them clearly. To enter $^{-}17$, for example, it is usual to first enter 17 and then to change its sign by pressing a 'change of sign' button, typically marked '$^{+}/_{-}$'. The calculators should also be able to add, subtract, multiply and divide directed numbers correctly. With all of these facilities present, a calculator becomes a powerful ally. As a simple introductory example of the need for negative numbers, one might consider, for example, an overnight fall of 7°C from an initial temperature of 5°C as recorded on a thermometer and mimic the result by working out $5 - 7$, using a calculator. Later, whatever means of explanation is used by an individual teacher, some pupils will need to be left with the conviction that a negative number multiplied by a negative number always results in a positive number. The calculator scores heavily here! Indeed, with suitable calculators available, it is possible for most pupils to discover all they need to know about the various operations on directed numbers simply through using a calculator.

Multiplication of Decimals

The inspection of answers obtained with a calculator can be used as a very effective introduction to the traditional pencil-and-paper rule. In the first stage of this approach calculators are used to work out ten to twenty examples of the following type: 0.9×0.7, 3.4×0.6,

2.06 × 4.7, 0.48 × 0.23, . . . The observation is made that, in each case, the number of digits to the right of the decimal point in the product is equal to the total number of such digits in the two numbers which have been multiplied. The second stage is to work out, without using a calculator, the same ten to twenty examples devoid of all decimal points, i.e. 9 × 7, 34 × 6, 206 × 47, 48 × 23, . . . The further observation is that the new results produce the digits in the answers to the original ten to twenty examples but give no clue as to the whereabouts of the decimal point. The two stages, taken together, give the basis for the rule: multiply the numbers, using pencil-and-paper, as if they were whole numbers and insert the decimal point in the answer so that the number of decimal places is equal to the total number of decimal places in the original numbers. In the original ten to twenty examples chosen, it is important to avoid examples of the types typified by 1.4 × 0.35 and by 1.60 × 0.7. The first type is characterised by having an even digit (in this case 4) coupled with the digit 5 in the least significant places of the two numbers. The second type has a trailing zero in one of the numbers. Both types suffer from the propensity of calculators to suppress trailing zeros. Thus the first example would normally give 0.49 (and not 0.490) when using a calculator and the second example would give 1.12 (and not 1.120).

Division by a Decimal

A systematic approach, using the observation of answers obtained with a calculator, can enhance pupil understanding and memory of the traditional pencil-and-paper algorithm. For example, consider the following batch of divisions, where the aim is to find a rule by which the answer to the first division can be carried out using pencil-and-paper: 2.884 ÷ 0.07, 28.84 ÷ 0.7, 288.4 ÷ 7. The first observation to make is that the last of the batch involves division by a whole number. It is assumed that this process can already be carried out efficiently by the pupils concerned. The second observation is that each of the batch gives the same answer when worked out with a calculator. The third observation is that one can pass from the first of the divisions to the second, and from the second to the third, by multiplying each number by ten. By pooling these observations and considering more batches of divisions, pupils can arrive at the rule: multiply each number by ten, and continue to do so until a division by a whole number is produced – then work that division out.

Even with the predominance of calculators as a means of carrying out divisions by a decimal, it may be useful to retain some element of the traditional rule to help in making rough estimates of answers.

Area, Volume and Capacity

There is no such thing as an awkward number when using a calculator. This means that the ideas of area, volume and capacity can be developed in conjunction with practical work, involving the actual measurement of the linear dimensions of real objects and containers. Such practical work gives an opportunity for pupils to gain an earlier appreciation of the ways in which the accuracy of a calculated result depends upon the natural errors in the data provided.

Percentages

The presence of a '%' key on so many calculators is a response to a demand from users in business and commerce who perhaps have not gleaned as good an understanding of percentages from their own schooling as they might! A calculator chosen for school use may benefit from the omission of such a key, so that attention can be focused, for example, on the interchangeability of 0.72, 72% and $^{72}/_{100}$ and the need to calculate 47×0.72 when finding 72% of £47.

Square Roots

A calculator for school use should have a square root key but it may be better with some pupils to ignore such a key during the introduction of square roots; much can be learned about the nature of square roots by first involving pupils in a system of trial-and-error using decimal search.

Drawing Graphs

Even the simplest quadratic function, $y = x^2$, is often drawn badly by pupils when first meeting it. This need not be so if a calculator is used to help produce extra points, with co-ordinates such as (0.2, 0.04), (0.5, 0.25) and ($^-$2.3, 5.29), which help to give a smooth graph.

Ratio

The very useful forms 1 : n and n : 1 become more readily accessible with calculators present. Simple examples can emanate from the pupils themselves, e.g. George has 34p and Jean has 47p; this gives a ratio of 34:47, i.e. ~ 0.72 : 1 or 1 : ~ 1.38 with the help of a calculator. These results are one step removed from the statements: 'George has about 28% less than Jean' and 'Jean has approximately 38% more than George.'

Trigonometry

The sine, cosine and tangent of an angle can be introduced through practical work, involving pupils in the measurement of the sides of a set of similar right angled triangles and the subsequent calculation of the relevant ratios of the side lengths. Many teachers used this type of introduction before the advent of calculators but had to contrive the situations so that the calculations of the ratios depended upon division by simple whole numbers; with a calculator, division by 12.3 is no more difficult than division by 12 and, in consequence, there is no longer any need to predetermine side lengths. The development of trigonometry becomes much easier with calculators. It also becomes approachable earlier (no need to wait for the mastering of logarithmic skills!) and to pupils of a wider range of ability than before.

Lengths, Areas and Volumes of Similar Objects

There are many familiar everyday objects which, at first sight, seem apt for an experimental approach to the ideas involved here, e.g. a set of three different sized shampoo bottles, of the same make. It is, in fact, quite difficult to find sets of objects which really are mathematically similar. This becomes apparent by using division to compare corresponding lengths on objects which are suspected of being similar. The comparison is easily possible with a calculator and yet is unlikely to be attempted without one! In this case the calculator becomes the catalyst for a worthwhile practical lesson in the classroom in which it is shown that the eye can deceive.

7.4 Choosing a Calculator

The successful implementation of CALIM in the classroom ideally calls for each pupil to have the individual use of a calculator during a given lesson. Also, there are times when it is highly desirable that all the calculators in use are identical. Thus the purchase of a set of calculators by the school becomes a practical necessity. The following specification sets out some of the features to look for in calculators intended for use with children, of a wide spectrum of ability, aged between eleven and sixteen.

Specification Details

(1) 8 digits LCD (Liquid Crystal Display).
(2) Automatic power saving cut off.

Figure 7.5

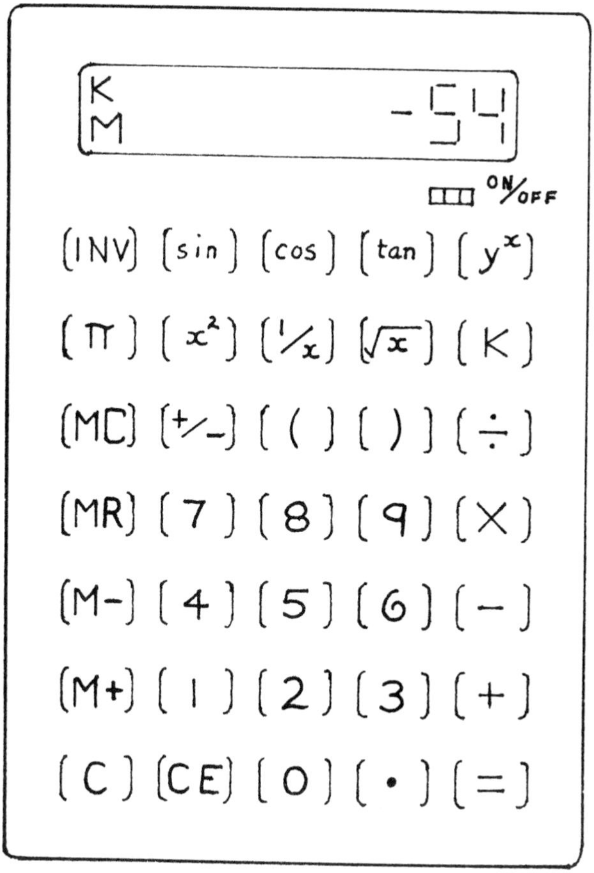

(3) Internal accuracy of operation to 9 digits with last digit truncated to give 8 digits displayed.

(4) Separate keys for clearing a number just entered 'CE' and for clearing a whole calculation 'C'.

(5) Change of sign key – negative indicator in correct position in display (see Figure 7.5).

(6) Brackets – perhaps as many as 6 levels of parentheses.

(7) No built in hierarchy for +, −, ×, ÷.

(8) Four key memory for: addition to memory, subtraction from memory, recall of memory and clearing of memory. Memory

indicator in display (see Figure 7.5).

(9) Separate keys for π, x^2, $1/x$ and $\sqrt{x}$.

(10) Constant facility for +, −, ×, ÷ by use of K key. Ability to set up as a 'Multiply repeatedly by 2' machine, for example, by using the key sequence '×', '2', 'K'. Constant indicator in display (see Figure 7.5). Cancellation of constant by second stroke of the K key or by any use of the keys for +, −, ×, ÷.

(11) Separate keys for sin, cos, tan and for finding powers (y^x key).

(12) INV key for finding roots and the inverses of sin, cos and tan, when used in combination with the keys for y^x, sin, cos and tan respectively.

(13) Well spaced keys with a positive action.

Notes Relating to Specification Details

(1) and (2) LCD and automatic power saving are vital for prolonged battery life and easy classroom management – in order to check that nimble fingers have not removed the batteries it is wise to collect calculators when switched on and then rely on the automatic switch off!

(3) Truncation: 1 ÷ 6 should yield 0.1666666 and not 0.1666667. The latter, while being a more accurate result, destroys pattern and can lead to confusion. For results out of range of the display, it would be welcome to find a calculator which displayed 'too big' or 'too small' rather than just an error message or zeros.

(4) It can be confusing to have a key which needs two presses to clear a calculation completely.

(5) The negative indicator should be to the immediate left of the digits in the display – not to the right of the digits or to the extreme left of the display as with some calculators.

(6) and (7) It is easier to follow the course of a calculation in the display if there are no built in priorities in the operation of the arithmetic keys, e.g. $2 + 3 \times 4$ should yield 20 and not 14. The presence of brackets allows the user to choose whatever hierarchy is needed, as the occasion arises.

(8) The presence of a combined memory recall and memory clear key on so many calculators is to be regretted – it means that the memory cannot be cleared without completely interrupting the course of a calculation.

(9) The keys for x^2, $1/x$ and $\sqrt{x}$ are immensely useful. The key for π could be replaced by an engraving of a value for it, elsewhere on the calculator, without any great loss.

(10) The switchable constant for +, −, ×, ÷ is a most useful feature for building up number patterns; it can help with multiplication tables at a simple level and compound interest at a more advanced level, to name but two applications among many.
(11) and (12) The keys for sin, cos and tan should work on the assumption that the number entered is an angle measured in degrees. It is unlikely that powers and roots of negative numbers will be achievable with the y^x and INV keys but it is worth looking out for innovations in this direction.
(13) Keys held in by an integral cantilever plastic 'spring' are to be avoided – they are not likely to last for much more than a year under heavy classroom usage. Keys which float freely but are held captive by the keyboard are more likely to give lasting performance.

8 GAMES AND RECREATIONS IN THE MATHEMATICS CLASSROOM

John Dunford

8.1 Why use Games and Recreations?

When considering how to teach a new topic few teachers will turn to games and recreations as a learning aid. HM Inspectors noted recently that the extent to which puzzles and games were put to mathematical use was 'disappointingly low'. Yet there are many books of puzzles and games and most mathematics textbooks have a section on these. Such sections are called 'Interlude' or 'Diversions' and occur either at the end of the textbook or at regular intervals in order to provide an occasional relaxation as a class works through the book. Little attempt is made to integrate these activities into the mainstream text and there is no advice on how the teacher may best use them. Many are used simply as 'fillers' for the last week of term or are given to a small group when the rest of the class has measles or is undertaking a geography field-trip. Yet, carefully designed games, played by a class in a well-organised manner, can enhance the learning experience in a number of ways.

Our general objectives may be summarised as (a) enjoyment, (b) increasing understanding and (c) broadening the scope of the subject. Enjoyment is our first and foremost objective. The inclusion of a game, puzzle or recreation must brighten the lesson and help the pupils – however good or bad they may be at mathematics – to look forward to the next mathematics lesson. In the hands of an able teacher, mathematics need never be a dull subject and the structured use of games can help all teachers to encourage the pupils and enliven the subject matter. This is particularly so for pupils who find mathematics difficult and who therefore have little to which to look forward in the next mathematics lesson. Pupils who find mathematics easy can also be bored by it and there are many activities which will increase their enjoyment of the subject.

Our second general objective is to use games and recreations in such a way that they deepen the understanding of a concept or create a new way of practising a skill. Games can be used to help in the teaching of many topics – four rules of number, co-ordinates, vectors, fractions, angles, bearings, to name a few. Indeed, if one finds that the games described below are helpful, it is not a difficult matter to develop one's own games in order to teach a particular topic. Lack of time, not lack of inspiration, is the greatest restraint for the busy teacher.

Thirdly, games may be used to broaden the subject beyond the boundaries of the syllabus. They are particularly useful for stimulating the most able pupils and, more generally, may be used to compensate for the deficiencies of examination syllabuses which are too restricted in their approach.

8.2 Games in Mathematics

Games for Learning Tables

Learning multiplication tables is not usually an enjoyable pastime, especially for twelve and thirteen-year-old children who have been trying to learn them for several years. Yet if the teacher regards them as essential – and it is undeniably useful for most children to know that $7 \times 6 = 42$ without spending a long time recalling the fact – then there are many games that can be played which will make the activity both enjoyable and rewarding. Bingo and dice games are two examples.

Bingo cards, prepared beforehand, may be issued to the class, with twelve counters each for covering the numbers. The teacher then calls, at reasonable speed, 5×6, 7×4, etc. and records the names on the blackboard of the first pupil to complete a row, a column, four corners and a full house. After each game, the cards are passed on. To avoid the need for counters, pupils can draw out and complete the bingo card with their own numbers, deleting them as they are called.

Figure 8.1

8	21	36	49
14	25	40	56
18	27	42	72

Dice, especially eight- or ten-sided, can be useful for multiplication games. Two dice are issued to each pair of pupils who take turns to throw the dice simultaneously and multiply the numbers. The results are added to the player's previous total and the winner is the first to reach, say, 500. To incorporate subtraction practice, a form of Dice Darts may be played, starting with 501, the winner being the first to reach zero. Refinements such as the need to start or finish with a double, or to finish with exactly 500 or zero, are best avoided since they do not contribute either to the enjoyment or to the main purpose of the game.

Inventing Games

Inventing a simple game to teach a particular concept or practise a skill is not as difficult as it sounds. A teacher who wanted to give a small class of less able pupils an exercise on the 24-hour clock hit on the idea of making a set of dominoes. Using felt-tip pen on blank playing cards (readily obtainable from manufacturers at a reasonable price per thousand), she soon wrote out a game in which 12-hour times had to be matched to 24-hour times. The pupils took turns to start (with any card) and the first one to put down all his cards was the winner. (N.B. all the cards must be dealt out.)

Figure 8.2

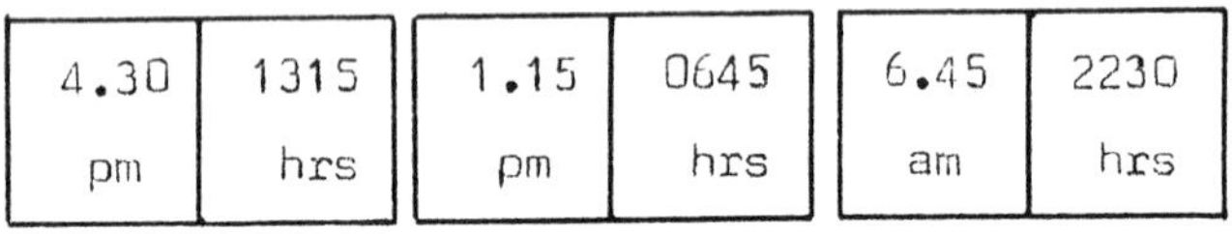

4.30 pm	1315 hrs	1.15 pm	0645 hrs	6.45 am	2230 hrs

Vector Games

Another teacher invented Vector Tennis which was used successfully to teach the meaning of vectors to a class of less able pupils, but which also appealed to some of my more able pupils. The two players are each

Figure 8.3

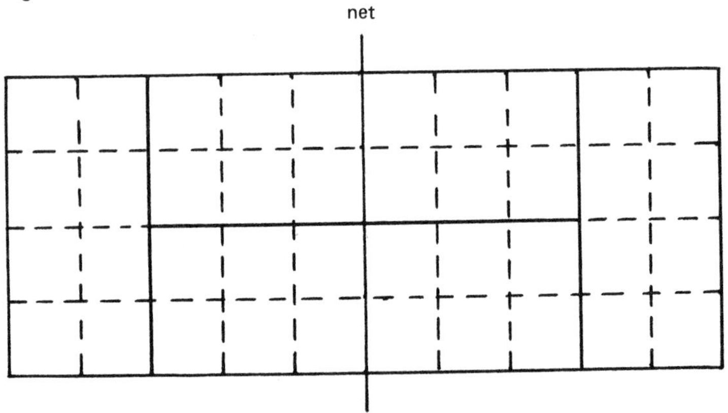

dealt five cards, on which are written vectors such as

$$\begin{pmatrix}3\\2\end{pmatrix}\ \begin{pmatrix}-1\\0\end{pmatrix}\ \begin{pmatrix}6\\-2\end{pmatrix}.$$

The ball (a counter) is 'served' into the opponent's court simply by placing it in the correct quarter. Cards are then played alternately until a player has no card which will return the ball into his opponent's side of the court. This player loses the point. A refinement is to have a number of 'Chance' cards which, if held, may be played at any time. The 'Chance' pile will contain cards which tell the player such things as 'Racket breaks. Lose point'. The 'rallies' can last for some time and, rather than use the tennis scoring system, the winner is the first player to reach an agreed number of points.

A vector game which appeals to children of all ages and abilities is Grand Prix. Several versions of this exist, including one that appeared in 'Mathematics in School' (November 1974, p. 24), but a version that is suitable for almost all pupils is the one shown in Figure 8.4.

(a) Two to four players may race on the same track.
(b) Each player chooses a position on the starting line.
(c) Each player starts by moving along vector

$$\begin{pmatrix}0\\+1\end{pmatrix}.$$

(d) Players then accelerate with moves

$$\begin{pmatrix}0\\+2\end{pmatrix},\ \begin{pmatrix}0\\+3\end{pmatrix},$$

etc., until they wish to slow down the forward movement and/or begin to move to the right.

(e) In order to have realistically gradual acceleration and deceleration, each component may be altered by +1, −1 or remain the same. For example, if a player's previous move was

$$\begin{pmatrix}0\\+3\end{pmatrix},$$

then he has a choice of nine moves for his next turn:

Figure 8.4

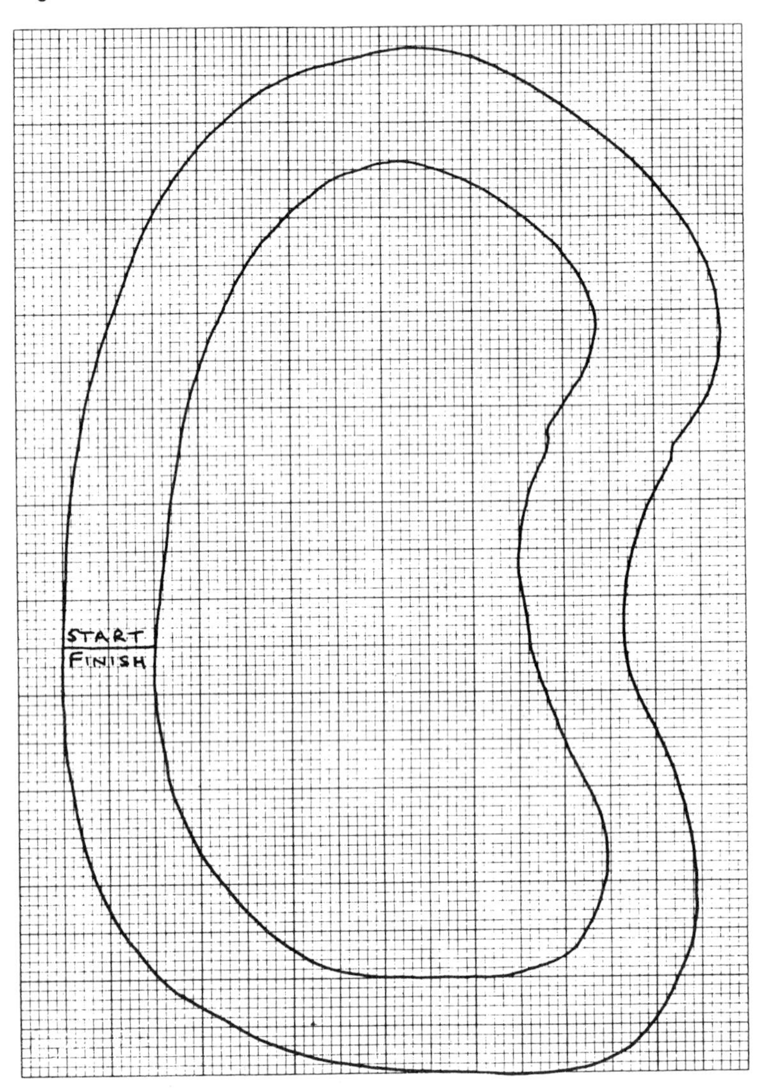

$$\begin{pmatrix}0\\+2\end{pmatrix}\begin{pmatrix}0\\+3\end{pmatrix}\begin{pmatrix}0\\+4\end{pmatrix}\begin{pmatrix}+1\\+2\end{pmatrix}\begin{pmatrix}+1\\+3\end{pmatrix}\begin{pmatrix}+1\\+4\end{pmatrix}\begin{pmatrix}-1\\+2\end{pmatrix}\begin{pmatrix}-1\\+3\end{pmatrix}\begin{pmatrix}-1\\+4\end{pmatrix}.$$

(f) A player may not land on a square already occupied by another player.

(g) If a player's car leaves the track, he loses two turns and starts again at the point where he left the track, as if his previous move had been

$$\begin{pmatrix}0\\0\end{pmatrix}.$$

(h) Pre-drawn tracks of varying difficulties (e.g. Figure 8.4) may be given to pupils at first, but they will soon be asking for blank graph paper on which they will draw tracks with more bends than Brands Hatch.

Games for Learning about Co-ordinates

If a game is being used as a teaching tool, then care must be taken to ensure that there is nothing in the game that will mislead the pupils. An example of this is the use of the Battleships game in SMP Book A (1970, p. 14). The pupils have to place their ships in the squares and then attack the enemy ships by firing at locations such as (3,4). Because the ships are situated inside the squares, (3,4) is not a point, but an area. This has applications such as map-reading but, in the next

Figure 8.5

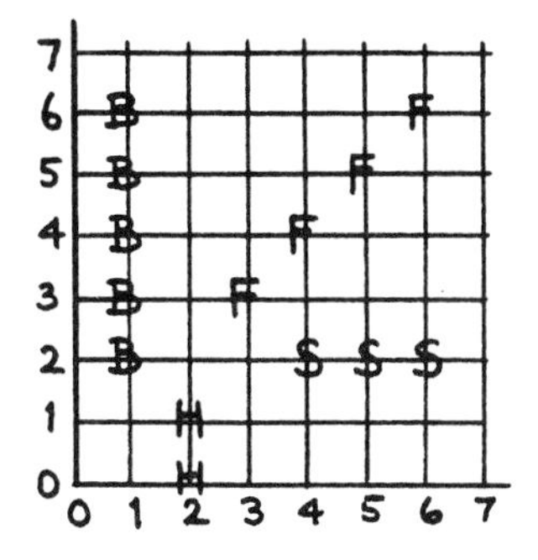

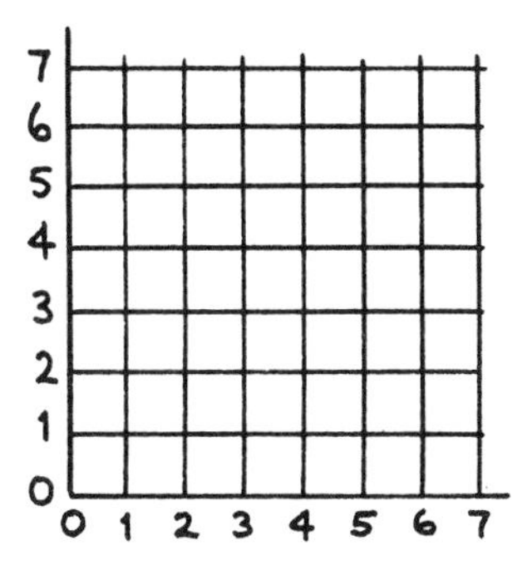

section of work, one has to think of (3,4) as a point and the game of Battleships is a poor preparation for this.

In order to overcome this difficulty, a colleague devised a more useful version of Battleships, which he called Air Raid, in which the aeroplanes are placed at the intersection of the lines. In this game, (3,4) identifies a point and the game is a most effective device in the teaching of co-ordinates. Each player marks on his co-ordinate grid five bombers, four fighters, three seaplanes and two helicopters and each of these four sets of planes must be placed in a straight line (horizontal, vertical or diagonal), as shown in Figure 8.5. In order to save time, prepared sheets are issued in pairs to each pupil, their own planes being placed on the left-hand diagram and their shots at the enemy being recorded on the right.

Another game for two players, described by Banwell *et al.* (1972, p. 79), which can be used to teach co-ordinates is called Four-in-a-line. The players take it in turns to say a pair of numbers and a third person, the marker, puts a point on the co-ordinate grid in the player's colour. The winner is the first person to have four points in a line (horizontal, vertical or diagonal).

Card Games

Unfortunately many children do not play cards and therefore the use of orthodox card games is greatly reduced. If, however, some pupils do enjoy a game of cards, then it is useful to have some packs made up in the following ways. Each pack has 52 cards, in four suits (two black, two red).

Pack 1: Equivalent fractions, the four suits being

$$\frac{1}{2}, \frac{2}{4}, \frac{3}{6} \ldots \frac{13}{26}$$

$$\frac{1}{3} \quad \ldots \quad \frac{13}{39}$$

$$\frac{1}{4} \quad \ldots \quad \frac{13}{52}$$

$$\frac{1}{5} \quad \ldots \quad \frac{13}{65}$$

Pack 2: Number bases, the four suits being

1_2, 10_2, 11_2, ... 1101_2
1_3, 2_3, 10_3, ... 111_3
1_4, ... 31_4
1_5, ... 23_5

Pack 3: Multiplication tables, the four suits being

3, 6, 9, ... 39
4, ... 52
5, ... 65
6, ... 78

Packs of cards such as these may be used for any card game that may be played with an ordinary pack – whist, bridge, snap, cheat, pelmanism, patience, etc.!

Games for Angles and Bearings

The major problem in the early stages of teaching angles and bearings is often that the pupils do not mentally relate the number of degrees with the physical size of the angle. Thus, given a protractor and an angle to measure, answers of, say, 65° and 115° are interchanged. One way in which to overcome this is to begin this section of work by playing Guess the Angle, a simple but enjoyable class game. An angle (or a bearing) is drawn on the blackboard and the pupils write down their guess of its size – three points for a correct guess, one point for being within 5°. Scores are often surprisingly high.

Of all the games that I have used the most popular has been a Bearings game called 'Rambler' which was invented by Harry Jeffery, a gifted teacher of mathematics in County Durham. The board is drawn (see Figure 8.6) on a large piece of card. A secret destination card is dealt to each of four players and the object of the game is to be the first to reach one's secret destination. Playing cards, which have been prepared with distances and bearings such as

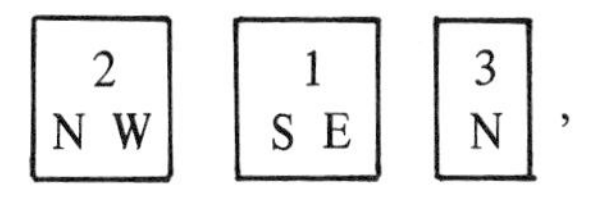

are dealt, five to each player. Cards are played in turn, an extra card being picked up to replace the one played, and the player moves his

Figure 8.6

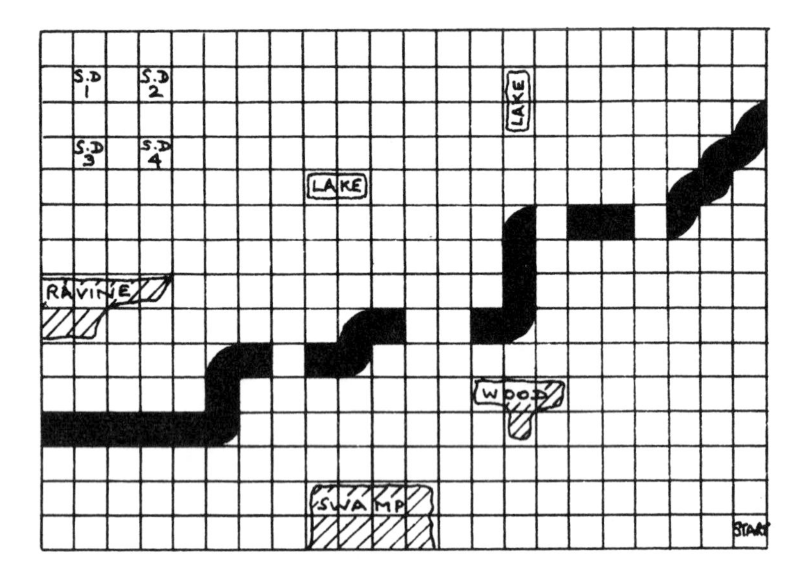

counter the appropriate length and direction. If a counter lands on a hazard, the player misses two turns; if it lands in the river, he returns to the start. The river may only be crossed at one of the bridges. A refinement of the game, which made it so popular with my pupils, is that some of the cards are red and, when a player puts down a red card, all players have to carry out the same move. The frequency with which players are forced into the river on an opponent's red card increases with practice! A further refinement is to have 'Take a chance' squares and a pile of chance cards.

A Function Game

Banwell *et al.* (1972, p. 90) describe a splendid Function game which is to be played in complete silence. Pupils are invited to write numbers on the blackboard and the teacher, having a function in mind, such as $f{:}x \to x + 4$, writes a number corresponding to each of the pupils' numbers

Pupil numbers	Teacher numbers	Space
2	6	
5	9	
⋮	⋮	

A space is left on one side of the blackboard for pupils to write comments concerning the function or, later in the learning process, to write down an algebraic statement of it.

When the pupils have grasped the essentials of the game, the teacher may introduce variants. After three pairs of numbers have been written down in a new game, the teacher may offer the chalk to the pupils to write the corresponding number in the teacher's column. Alternatively, the teacher may write a number initially in the second column and ask the pupils to write the appropriate number in the first column.

This game may be used at any stage in the teaching of functions and, if played in complete silence, can create a most attentive atmosphere. Devices such as this should not be used too often, but the variety which is given to the teaching process can bring great benefits.

Games and Generalisations

There are many games which do not fall into categories such as number games, angle games or vector games, which one might use as a teaching aid at a certain point in the syllabus. Nevertheless they are well worth playing in the context of a mathematics lesson or club. The first group of these games is a selection of those for one player from which algebraic generalisation may be drawn.

(1) Tower of Hanoi

The game may be easily constructed with three pieces of dowel rod set in a wooden board and circles of card of decreasing size with holes in the middle to fit over the dowel. The object of the game is for the player to move the circular discs from one pole to another, one at a time, so that the discs finish in the same order on another pole. No disc may be placed on top of a smaller disc. This puzzle has a long history, but pupils new to it should start with two discs and record how many moves are made in order to obtain the required result. After carrying out the sequence for two, three and four discs, the pupil should be invited to generalise the result, which is that $2^n - 1$ moves are required.

Figure 8.7

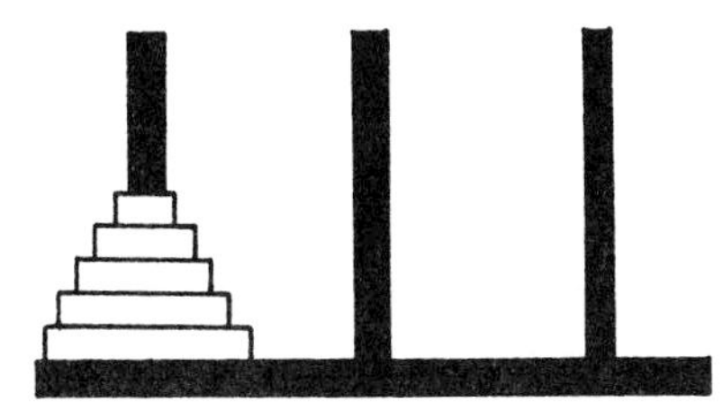

(2) The Lucas Problem (otherwise called Frogs)

Figure 8.8

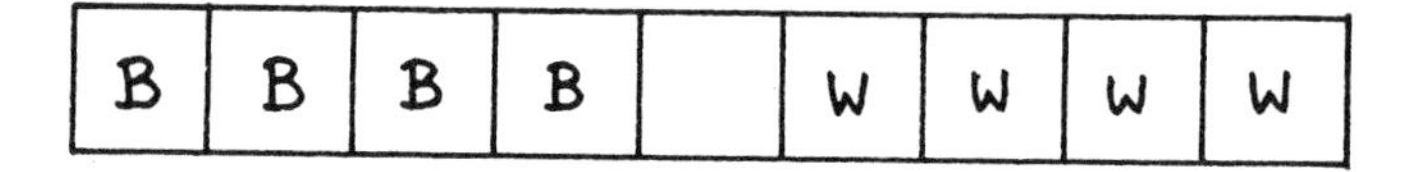

Four black and four white counters are placed in a line as shown in Figure 8.8. The object of the game is to interchange the positions of the black and white counters. Moves may be made either by sliding a counter into the adjacent vacant space or by jumping over a counter of the opposite colour into the vacant space. Black counters may only be moved to the right and white counters to the left. This is considerably more difficult than it looks and, if success is not achieved, the number of counters may be reduced. With one counter on each side, and then two counters on each side, the problem is much easier. The number of moves required should be recorded and the pupil invited to generalise the result, which is that $(n + 1)^2 - 1$ moves are needed.

(3) Intersecting squares (Figure 8.9)

Figure 8.9

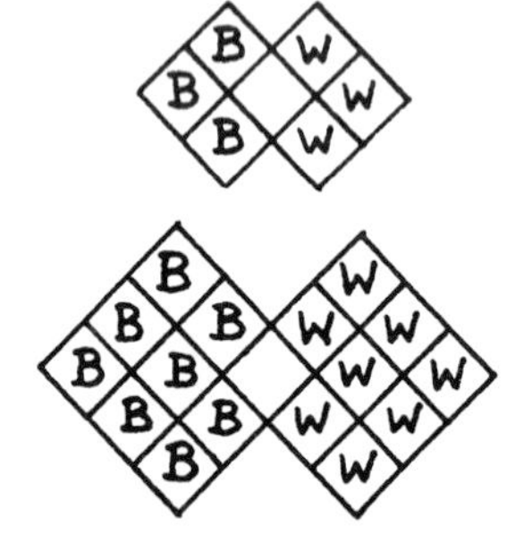

As in the Lucas Problem the object of the game is to exchange the black and white counters in the minimum number of moves by sliding and jumping. When a strategy has been devised using the smaller board, then the larger board may be tackled. Enthusiasts will design a board with 15 counters on each side and attempt to generalise the relationship between the number of counters and the minimum number of moves.

Other Strategy Games

There are many games for two players which require considerable thought and for which winning strategies may be devised. Some which have been found to appeal to older and younger pupils alike include the following:

(1) Nim

Figure 8.10

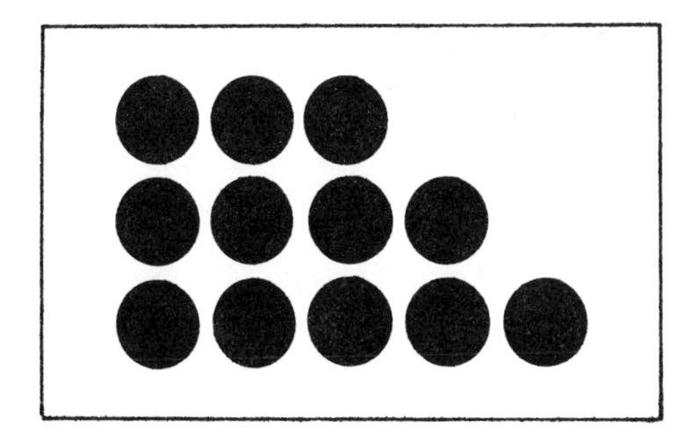

Twelve counters are arranged as shown in Figure 8.10. Taking turns, the two players pick up one or more counters each turn, never taking from more than one horizontal row at a time. The player who collects the last counter is the loser. As a variation the players may agree that the one who picks up the last counter is the winner.

The game may be played with any number of counters in any number of rows and a winning strategy has been devised for this game using binary numbers. This is fully explained in Martin Gardner's first book of *Mathematical Puzzles and Diversions* (1959, Penguin, p. 154).

(2) Ticktacktoe (Figure 8.11)

Figure 8.11

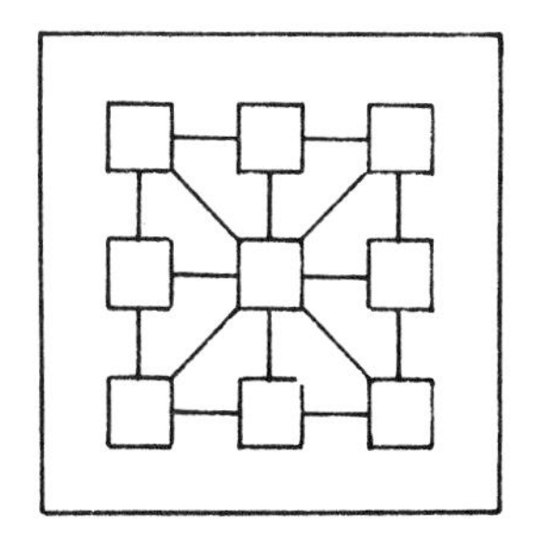

Two players have three counters each which they place, in turn, on to one of the nine spaces. The object of the game, like noughts and crosses, is to get your three counters in a line. If this has not occurred by the time all six counters are on the board, play continues with players moving, in turn, one counter at a time along a line to an adjacent space. Whether I took first or second turn I was consistently beaten at this game by an eleven-year-old pupil who was eventually kind enough to explain his strategy to me! A fuller discussion of this game, and its variants, occurs in Gardner (1959, p. 44).

(3) Hex (Figure 8.12)

Figure 8.12

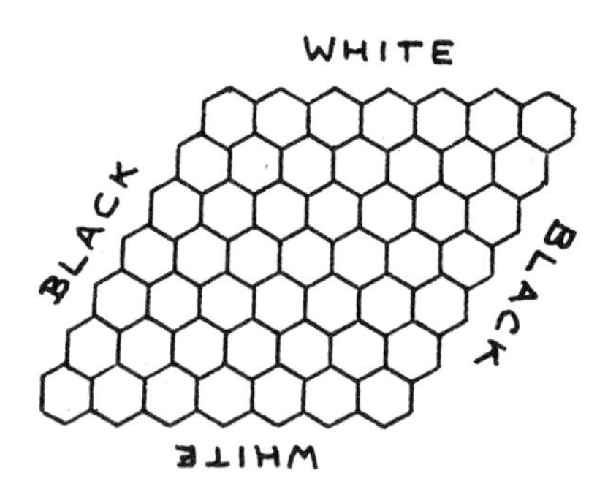

The board is diamond-shaped, each side containing any number of hexagons from six to twelve, eleven being the number most often used. A lot of counters in each of two colours are needed. The two players take turns to place a counter on the board, the winner being the first to make a continuous path of his counters join the two opposite sides that are in his colour. Counters, once placed, may not be moved. Gardner gives an interesting account of the origin of this game, which was invented as recently as 1942, and discusses the strategy for winning (1959, p. 70).

(4) Nine Men's Morris (Figure 8.13)

Figure 8.13

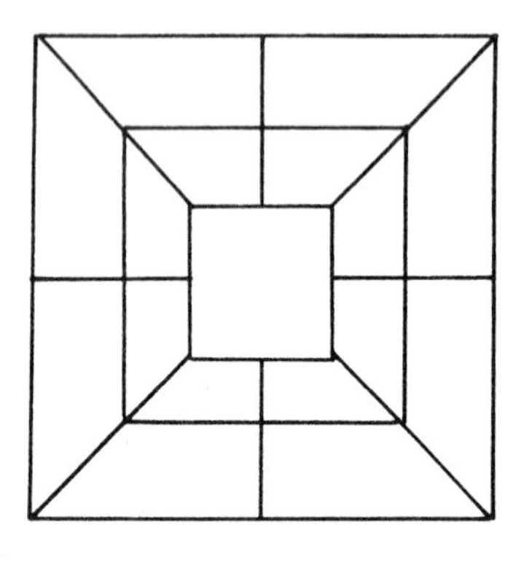

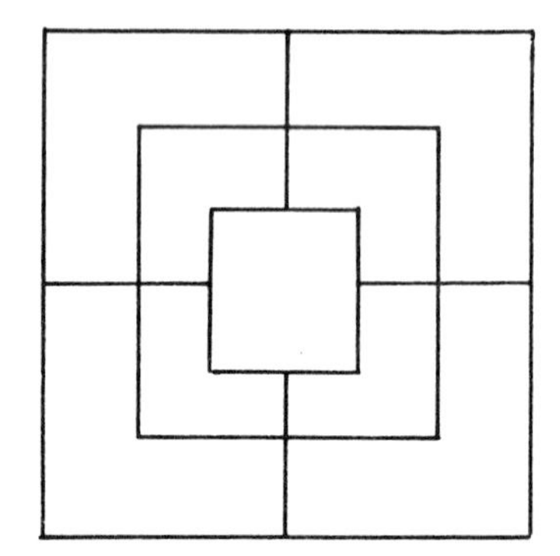

Two alternative boards are given in Figure 8.13 for the game of Nine Men's Morris, an ancient game with several variations. The two players have nine counters each, which they place in turn at one of the corners or intersections on the board. When all the pieces have been put down, players may move to an adjacent point, the object being to reduce the number of one's opponent's counters to two or else to prevent him from moving. When a player has three of his counters in a row he may take away any one of his opponent's counters. If he breaks his own row of three by moving one piece, the row must be restored in his next move.

8.3 Mathematical Puzzles

Books of mathematical puzzles are not hard to find. Martin Gardner's series of books is well known and Boris Kordemsky's *Moscow Puzzles* has proved addictive to mathematicians and non-mathematicians alike. Yet there are dangers in introducing pupils to puzzles in mathematics lessons. Many pupils, especially the less able, find them difficult and quickly become bored. The solution of certain types of puzzle too clearly defines the enjoyment to be gained from it. Furthermore, many puzzles can only be done once and so the teacher's time spent in preparing the puzzle for pupil use is not being spent efficiently. There are also puzzles which may lead to some discussion, but which unnecessarily confuse the pupil. Such puzzles usually contain a catch, such as the question in which a frog is at the bottom of a 10m well. Each hour it climbs up 1m and then slips back 0.5m. How many hours, the question asks, does it take the frog to climb out?

There are puzzles, however, in which the pupil can gain enjoyment from the pursuit of success as much as from its conclusion, where even a partial solution can give pleasure. The following are examples:

(1) Coloured Cubes

Figure 8.14

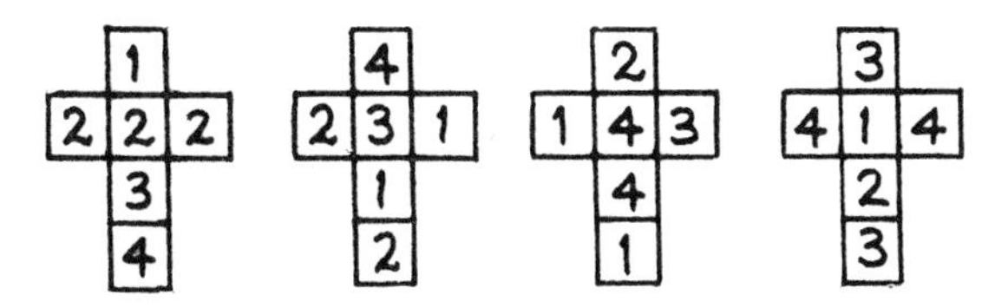

Figure 8.14 illustrates the nets of four cubes which can be made simply from card or wood and coloured according to the plan shown, each number representing a different colour. The aim is to arrange the cubes in a line in such a way that four different colours are exposed in each direction.

(2) Matchstick Puzzles

Figure 8.15

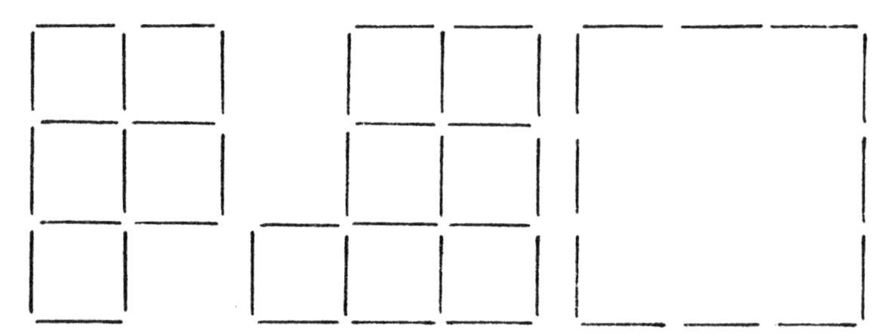

The first diagram in Figure 8.15 provides us with two puzzles. First, the pupil should remove three matches so that only three squares are left. Replacing these, he should then try taking away four matches so that only two squares remain. In the second diagram, three matchsticks must be re-arranged so that five squares remain. Finally, the three-by-three square provides us with two area problems. First, leaving the original twelve matches in the same position, the square should be divided into three equal parts by the addition of seven more matches (Pearcy & Lewis, 1966, p. 54). Then, the twelve matches should be moved so as to enclose an area of exactly four square units (Gardner, 1959, p. 101).

(3) Number Puzzles

(i) Cross-numbers

Like crosswords, these may be any size of square or rectangle and may be made up by the teacher as an interesting practice in any of the four rules. Blank cross-numbers may be given to the pupils for them to fill in and make up the clues.

(ii) Magic Squares

There are many informative accounts of the history and theory of magic squares. In general, an $n \times n$ square is filled with the consecutive numbers 1 to n^2 such that the numbers of every row, column and diagonal add up to $\frac{1}{2}n(n^2 + 1)$. Kordemsky (1975, p. 146) and Gardner (1961, p. 104) give some interesting variations and explain how to construct a magic square.

Many pupils will be able to construct the 3×3 magic square for themselves, although it is sometimes necessary to tell them to put the number 5 in the middle. With 4×4 magic squares it is useful to have some in one's desk where a few of the numbers have been filled in.

(iii) The Four Fours

'How many numbers can you make using four fours?' is a question that, with a few examples and a hint or two, has stimulated many of my pupils to hours of mathematical activity. The idea is a simple one, but the quest for solutions is fascinating and there is a considerable literature on the subject ('Mathematics in School', July 1973, p. 28). Some numbers, such as $5 = (4 \times 4 + 4)/4$, may be expressed using only the integer 4 and the symbols of the four rules.

Interested pupils, having reached their limit with the four fours, may find four threes or four nines a similar challenge.

(iv) Divisibility

Ask the members of a class to write down a three-figure number. Then ask them to make a six-figure number by repeating the three-figure number. Dividing this six-figure number by 7, 11 and then 13 should bring them to the original three-figure number. The division can be done correctly by most pupils but it is of more interest to ask them why this occurs – $7 \times 11 \times 13 = 1001$. Further interesting divisibility problems and puzzles are given in Kordemsky (1975, p. 134).

(v) Russian multiplication

Example: 35×18

Method:
(a) Multiply 35 by 2, divide 18 by 2.
(b) Ignore remainders.
(c) Continue until the number 1 appears in the division column.
(d) Delete all lines in which there is an even number in the division column.
(e) Add the numbers remaining in the multiplication column to obtain the result.

35	18
70	9
~~140~~	~~4~~
~~280~~	~~2~~
560	1

Answer 630

(vi) Chain puzzles

Method:
(a) Take any number.
(b) If it is even, divide it by 2.
(c) If not, multiply it by 3, then add 1.
(d) Continue the sequence, following these two rules.

(e) Investigate what happens.

Example: 18, 9, 28, 14, 7, 22, 11, . . .

There are many such chain puzzles; indeed, an armchair, the back of an envelope and a few minutes to spare will yield further chains. A second example is as follows:

Method: (a) Take any number.
(b) Multiply its digits together.
(c) Multiply the digits of your new number.
(d) Continue this until you have a single-digit number.
(e) How many steps did it take?
(f) Investigate this with other numbers.

(4) Puzzles for the Calculator

Given a calculator and no further instructions, most children will either press the keys aimlessly or turn it upside down and make words such as 'Shell oil'. To allow this to happen is to lose an opportunity to open young eyes to the beauty of mathematics and the following suggestions, written on cards and kept in the teacher's desk to be produced at the right time, will bring pleasure to many:

(i) $12 \times 42 = 21 \times 24$
$13 \times 62 = 31 \times 26$
$23 \times 96 = 32 \times 69$
Can you find other multiplications like this, where reversing the numbers yields the same result?

(ii) $9 + 9 = 18$ $\quad 9 \times 9 = 81$
$3 + 24 = 27$ $\quad 3 \times 24 = 72$
Can you find other pairs of numbers that do this?

(iii)

27	39	69
+72	+93	+96
99	132	165
	+231	+561
	363	726
		+627
		1353
		+3531
		4884

The object of this is to produce a palindromic number (a number that reads the same in both directions). Some numbers, like 89, take many steps to reach a palindrome while others, like 196, are believed to be impossible (Kordemsky, p. 174).

(iv) The following can be given to pupils to calculate and continue the sequences:

(a) $1 \times 9 + 2 =$
$12 \times 9 + 3 =$

(b) $12345679 \times 9 =$
$12345679 \times 18 =$

(c) $142857 \times 2 =$
$142857 \times 3 =$

(d) $143 \times 7 =$
$286 \times 7 =$
$429 \times 7 =$

(e) $15873 \times 7 =$
$31746 \times 7 =$
$47619 \times 7 =$

(f) $6 \times 7 =$
$66 \times 67 =$
$666 \times 667 =$

(5) The Folded Sheet (Figure 8.16)

One of Gardner's puzzles (1961, p. 122) is to divide a sheet of paper into eight equal squares and number them as shown. The sheet has to be folded so that the numbers appear, in sequence, on top of each other. To do the same thing with the second rectangle is much harder.

Figure 8.16

1	8	7	4
2	3	6	5

1	8	2	7
4	5	3	6

(6) Letters for Numbers

```
  SEVEN        CROSS        ONE
+EIGHT       +ROADS       +ONE
------       ------       ----
TWELVE       DANGER       TWO
------       ------       ----
```

In each of these puzzles – and there are many others like them – each letter stands for a number. The object is to find the numbers which make the sums 'work'. Many pupils will get pleasure from solving this type of puzzle, although it is often necessary to give them a start by putting in one of the numbers. No digit can be used to stand for more than one letter.

(7) Mathematical Fallacies

Pupils' enjoyment from mathematical fallacies often comes from proving to their parents such delights as 1 = 2 and then asking them to explain what has gone wrong:

$$a^2 - b^2 = (a + b)(a - b)$$
$$\text{Let } a = b\text{, then } b^2 = ab$$
$$a^2 - ab = (a + b)(a - b)$$
$$\text{i.e. } a(a - b) = (a + b)(a - b)$$
$$\text{Divide by } (a - b)\text{, then } a = a + b \qquad [\text{fallacy: divide by } a - b = 0]$$
$$\text{But } a = b\text{, so } a = 2a$$
$$\text{and } 1 = 2$$

(8) Number Detective Work

Ball and Coxeter (1974, p. 5) describe several ways in which numbers chosen by someone may be found by another person after a series of operations has been performed. There is no better illustration of the value of algebra than this game and it may be played at any time when algebra is being used. Without the algebraic explanation, it is a source of wonder to the person who chose the number that another person should be able to discover his secret; once the algebra has been mastered it is quite simple to devise new puzzles oneself. One example is this:

(a) Ask someone to think of a number (preferably fairly small) and then to
(b) Multiply it by 5,
(c) Add 6,

(d) Multiply the sum by 4,
(e) Add 9,
(f) Multiply the sum by 5.
(g) Ask the person to tell you the result, subtract from it 165, divide by 100 and the answer will be the number originally chosen.

The algebraic explanation of this is as follows:

(a) Let the number be n, then it becomes
(b) $5n$
(c) $5n + 6$
(d) $4(5n + 6) = 20n + 24$
(e) $20n + 33$
(f) $5(20n + 33) = 100n + 165$.

8.4 Other Activities

Problem Cards for the Most Able

A recurring problem for many teachers of mathematics is 'What do I do with X when he finishes early?' X may be one of the most able pupils in the school or a quick worker in a lower set and one way to overcome this problem is to keep a small file of cards, to which additions may be made at any time, containing puzzles, recreations, problems and open-ended questions. Cards may be sorted into several levels of difficulty and an appropriate one drawn out for the early finisher. Some examples of these are:

1. SEVENTHS

What is $1/7$ as a decimal?
After how many places does it recur?
Use your calculator to find $2/7$, $3/7$, etc.
Have you found a pattern?
Now try Card 2

Equipment needed: calculator

2. RECURRING DECIMALS

On Card 1 you investigated 7ths.
Use the same method to investigate
Ninths
Elevenths
Thirteenths
etc.
Equipment needed: calculator

3. EVEN NUMBERS

Choose any even number
Multiply it by the next even number
Multiply your answer by the next even number again
Do this for several chosen numbers
What is the largest number that divides exactly into all the answers?
Can you PROVE this number always divides into three successive even numbers multiplied together?

3. AREAS AND PERIMETERS

Under what conditions are the area and perimeter of a triangle numerically equal?
Can you extend this to other polygons?

If a series of magazines such as 'Mathematics in School' is kept in the classroom or departmental library, then the spirit of inquiry may be encouraged by cards such as

THE NUMBER 111

Find *Mathematics in School* (Nov. 1980) from the bookcase in Room 36.
Turn to p. 20 and look at Section 2, 'Number Scales'.
Use your calculator where necessary and write down your conclusions.

Some very interesting puzzles, often capable of solution by algebraic methods, occur in the Brain Teasers in the Sunday newspapers. These may be cut out and pasted onto card to join the series suggested above.

The Golden Section

The Golden Section, or Golden Ratio, should be on every mathematics syllabus and is a fascinating topic for recreational study. This may be

Figure 8.17

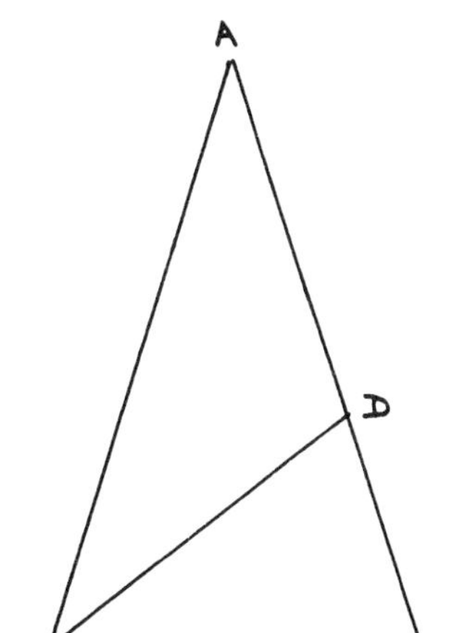

Figure 8.18

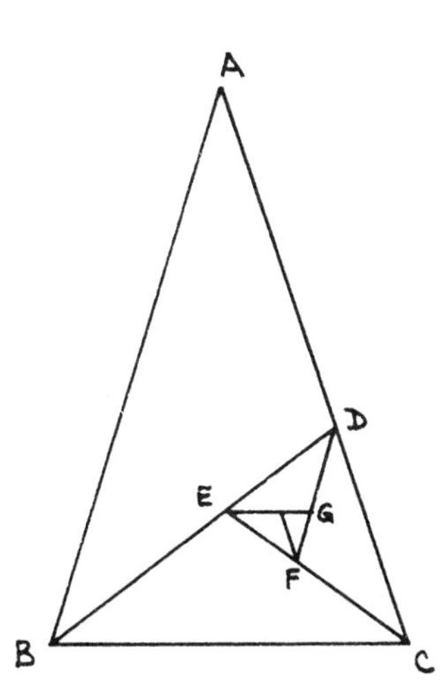

at a simple level such as in Figure 8.17 where an isosceles triangle is drawn with angles B and C equal to 72°. The ratio

$$\frac{AB}{BC} \simeq \frac{8}{5}.$$

If angle B is bisected, to meet AC at D, then

$$\frac{BC}{CD} \simeq \frac{8}{5}$$

and this process may be continued as in Figure 8.18, with each triangle having the long side divided by the short side equal to the Golden Ratio. The nest of triangles so formed resembles a snail's shell, a natural phenomenon in which is also found the Golden Ratio.

Pupils who wish to explore the exact value of the Golden Ratio will find it from the positive root of the quadratic equation $r^2 - r - 1 = 0$. Expressed as a decimal, this root is equal to 1.618033 . . ., the reciprocal of which is 0.618033 . . . Many other interesting properties of the Golden Ratio are discussed by Martin Gardner (1961, p. 69).

A starting point may be made with the Golden Rectangle. A rectangle WXYZ is first drawn with sides equal to those of the triangle in Figure 8.19. The second rectangle PQXY is obtained by defining the point P such that ZP = ZW. The process may be continued in the same way and the ratio of the sides calculated for each rectangle.

Figure 8.19

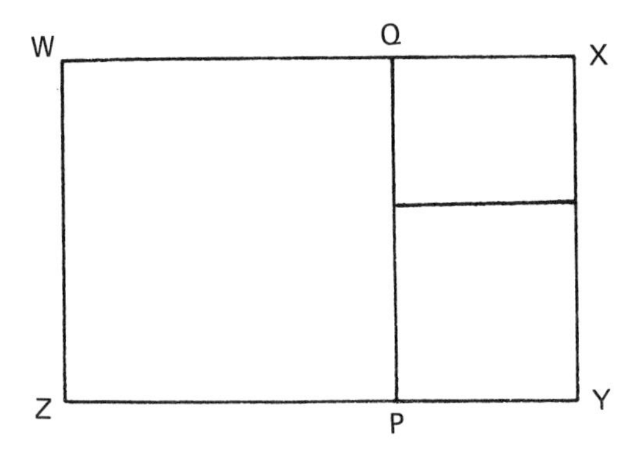

At a more practical level, the idea that the Golden Ratio is found in the shell of a snail may be extended to studies of fern and flower species, butterflies, even snowflakes – all these are outlined with beautiful simplicity in Fred Gettings (1963, p. 65). If you can obtain a copy of this book it is well worthwhile, for he also gives a fascinating account of the use of the Golden Ratio in art and architecture from the Temple of Diana at Ephesus, through Raphael and Leonardo, to Turner, Lowry and Mondrian. Pupils who become particularly interested in these links with art should be directed to Marilyn Lavin's book with its careful analysis of *The Flagellation* by Piero della Francesca (Lavin, 1972). Useful work may be done on this topic in conjunction with colleagues in the art and science departments.

Paper Folding, Origami, Tangrams

Given a sheet of paper no pupil need ever be bored if they have been introduced to the art of paper folding. Children of all ages can gain mathematical insight by obtaining a square, triangle, hexagon or, most satisfyingly simple of all, a pentagon from a sheet of paper. All that is needed for the pentagon is a strip of paper. This is then 'knotted' and the ends pulled tight. The result is illustrated in Figure 8.20. Clear descriptions of these foldings are given in Todd (1968, p. 139).

By using a triangle or quadrilateral in paper, cutting off the vertices and re-arranging them around a single point, pupils can quickly appreciate the sum of the angles of these figures and elementary geometrical constructions can equally well be introduced by paper folding (Brunton, 'Mathematics in School', July, 1973, p. 25).

Figure 8.20

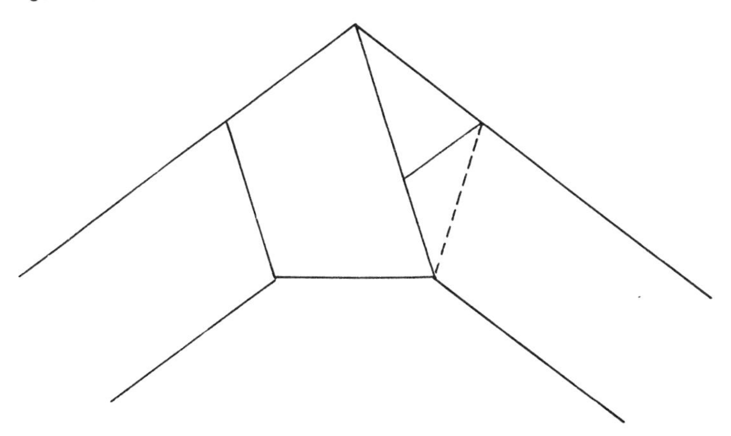

Figure 8.21

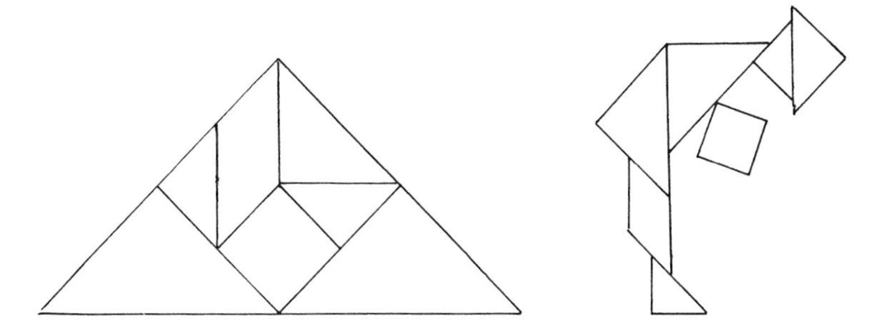

Another topic in which paper folding and cutting can be used as a teaching aid is Area – the triangle, trapezium and parallelogram are three examples of shapes where the area formulae can readily be demonstrated in this way.

Of less direct mathematical benefit, but of great interest to many pupils is the ancient Japanese art of Origami. Beautiful animal and human shapes are included in the enormous variety of designs on which several paperback books have been published.

The making of a pattern with the Chinese Tangram is a good geometrical exercise and, when cut out, the pupil can use the pieces to make many different designs (Figure 8.21). The story of Tanya Gram ('Mathematics in School', September 1981, p. 14) could be used as an interesting starting point for younger pupils.

Figure 8.22

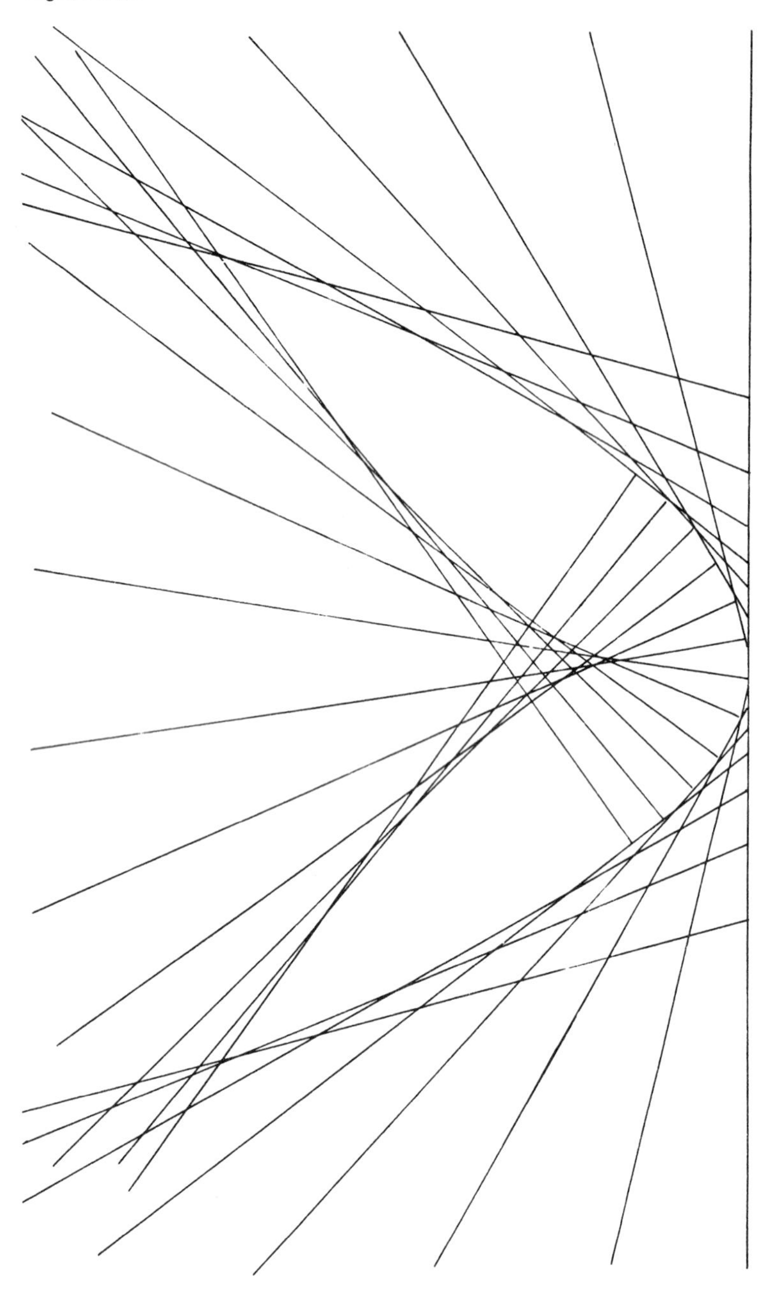

Curve Stiching and Drawing

Curve stitching appeals to pupils of all abilities and can be used to develop the understanding of, and skill in, several parts of the subject. The usual approach is to start with a simple astroid drawn on paper and introduce the pupils gradually to more intricate, decorative or mathematically difficult curves. The nephroid and cardioid are described in many places and Audrey Todd (1968, pp. 52, 56) also explains the Spiral of Archimedes and the Curve of Pursuit. From these drawings pupils can be encouraged to produce works which require considerable thought before putting pencil to paper and which bring mathematics and art closer together. Envelope curves can also be produced for the cardioid, nephroid, parabola (Figure 8.22), hyperbola, limaçon and deltoid (Lockwood, 1961); such work provides a stimulus for many pupils and the need for precision increases their skill in the use of geometrical instruments.

Three-dimensional curve stitching boards may be obtained commercially and add a sculptural aspect to the artistic work described above.

Construction of Solids

Figure 8.23

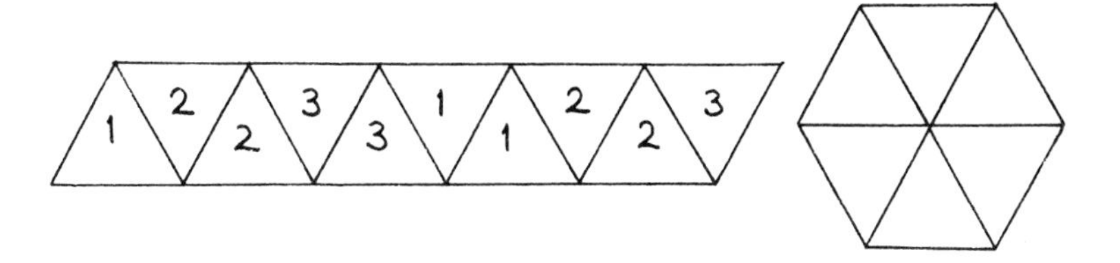

The drawing of nets for solids also provides good practice in the use of geometrical instruments and the Hexaflexagon, although not itself a solid, provides a good starting point. Ten equilateral triangles are drawn on thin card and numbered as shown in Figure 8.23. The card is cut out and scored on both sides and the reverse side is numbered as follows: blank, 3, 1, 1, 2, 2, 3, 3, 1, blank. If the line of equilateral triangles is then folded so that only the number 1 faces are uppermost, the two blank faces may be glued together. Turning over the hexaflexagon it will be found that all six triangles have the same number, the six triangles with the third number being hidden inside. In order to reveal these, the edges have to be turned and opened out.

The hexahexaflexagon is similarly constructed from a row of

Figure 8.24

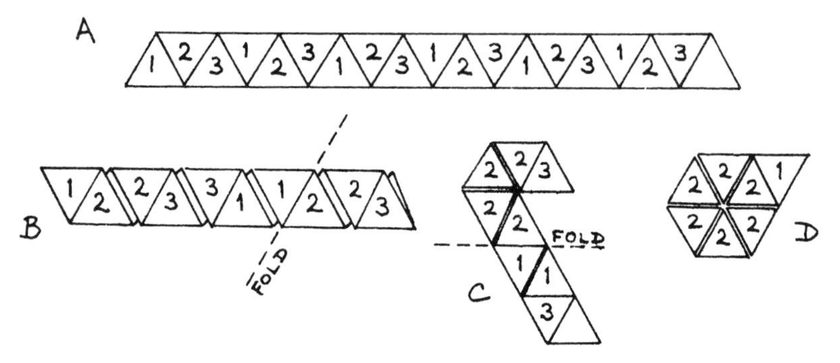

nineteen equilateral triangles which are numbered 1 to 3 on the front, as shown in Figure 8.24. The reverse side is numbered blank, 4, 4, 5, 5, 6, 6, 4, 4, 5, 5, 6, 6, 4, 4, 5, 5, 6, 6. The strip is first folded to the pattern in B, before being folded back along the dotted line shown. This yields the pattern C, which is again folded back along the dotted line in order to form the hexagon in D. The two blank triangles are then glued together. The two outside faces of the flexagon are now numbered 1 and 2 and the other numbered faces can be revealed by pinching two adjacent triangles together and pushing in the opposite vertex. Continuous flexing will bring up the faces numbered 1, 2 and 3 more often than those numbered 4, 5 and 6 (Gardner, 1959, p. 13).

The construction of solids has given hours of pleasure to both children and adults and their use as classroom decorations is evident in many lively mathematics departments. The benefits to pupils in increased geometrical skill and improved motivation more than repay the time involved in giving careful instructions at the outset; in particular, the problem of gluing the completed solid has to be faced. Only the best glue will suffice if a sticky mess is to be avoided. The two best books on the construction of solids (Cundy and Rollett, 1961; Wenninger, 1971) contain enough models, with instructions for nets, to occupy a pupil's spare time for several years. Yet one does not have to construct a Great Dodecahemicosahedron in order to derive pleasure or benefit from the exercise. Two of the mathematically least able pupils that I have ever taught were able to look with pride at their five regular Platonic Solids (tetrahedron, octahedron, cube, dodecahedron and icosahedron) which for many months remained on the classroom noticeboard.

Commercial Games

New commercial games are appearing so fast that it is impossible to include a comprehensive review here. Children are attracted to these for many reasons, not least because they seem even further removed from school work than something that has been prepared by the teacher. The higher quality of presentation of a commercial game greatly aids its impact. The games fall into three categories: those which have been produced specifically as a teaching aid, those in which the mathematical basis of the game can incidentally be put to use by a teacher and those where the mathematical content is small but which can nevertheless be used beneficially in club or classroom.

Into the first category come such games as *Decimal Dominoes*, *Taktiles*, *Centicubes*, *Pick a Pair* and *Dice and Dots*. The *Fletcher Maths* series is augmented by games such as *Race Track* and *Safari*, whilst the *Follow-up Maths* series is complemented by arithmetic games such as *Tablemaster* and *Add-Venture*. Macmillan's *Mathematical Games* is produced as a booklet of A2 sheets held together by a spine binder so that the games can be taken out and used separately. Many of these are simple arithmetic or logic games which would appeal to children of primary school age, but there are some – such as *Up and Down The Steps*, where positive and negative numbers are used – which are well suited to children at the younger end of the secondary-school age range. A Concept Chart for the games is included at the back of the booklet.

At the time of writing the second and third categories of games are enjoying a boom, thanks largely to the *Rubik Cube* and its related puzzles. The cube was invented by Professor Ernö Rubik, of the School for Commercial Artists in Budapest, in order to improve the three-dimensional spatial awareness of his students. Mathematical solutions to the problem of making each face into a single colour are based on group theory, although a knowledge of this branch of mathematics is not essential for its solution (as can be observed many times during a school day!) A contributory factor to the popularity of the *Rubik Cube* is undoubtedly its convenient size and shape – it can be carried in the pocket and played anywhere. *Scrabble* does not have this advantage, but its mathematical equivalent, *Equality*, could prove a useful aid to teachers. The tiles are either numerals or signs – brackets, equals, plus, minus, times, divide. The idea of the game is to form equations in crossword style, scoring points according to the numbers on the tiles which are put down. *Multipuzzle* has a tray, with squares six-by-ten, and 35 differently shaped hexominoes. The object of the game is to fill the tray completely with ten of the shapes. In the book which accompanies the game, a series of harder puzzles is given. Strategy games involving the arrangement of pieces on a board include *Othello* and *Reversi* where one's opponent's pieces may be trapped and turned over to become one's own. *Skirrid* is played with tiles of different shapes and the simple mathematics involved in calculating one's score can be used in the classroom to stimulate even the least well-motivated pupils. Three-dimensional *Noughts and Crosses* benefits from being quick to play and the perspex structure can be easily assembled.

The Maths Club

Apart from their uses in the classroom, games and recreations may be used to fulfil our general objectives in the more relaxed atmosphere of an extra-curricular Mathematics Club. Such a club is best started by a team of teachers, rather than by one individual. The preparation load is then spread more widely and the supervision of the club is improved. Good supervision is especially necessary because a well-patronised club is likely to take place in several rooms which may be used separately for games, practical activities, calculator work and number puzzles, for example. Once initial preparations have been made – and a combination of commercial and home-made activities can be assembled without too much trouble – the best way to start the club is not by an announcement in Assembly but to approach the pupils through their mathematics teachers, giving them an idea of what will be taking place. If the approach is right a high turnout is likely and so entry will have to be restricted to pupils of one age group at first. Older pupil volunteers will have to be enlisted to help to explain the rules of the games and activities for, given the basic equipment, children soon make up their own rules and, whilst this may be quite acceptable and may be socially beneficial, they may also benefit from abiding by the correct rules! Children do not play board games as much as their parents did and are therefore not accustomed to reading and interpreting a set of written rules. One cannot rely on writing down the rules on a card and leaving the children without further explanation to play the game. One problem here is created by the attractiveness of the Maths Club to the least able pupils. If they are to come along – and they are surely the pupils who should be most welcome at such a club – then they may need to be tactfully diverted from games or activities which they will not understand or else the spectre of failure, which haunts them in mathematics lessons, will soon discourage them from attending the club.

If the money allocated for mathematics will not stretch to the purchase of equipment for a club, then other sources of finance will have to be explored – the School Fund, the PTA or the Headmaster's generosity, for example. Finally, if there are no regulations to the contrary in your school, then a nominal charge can be made to the pupils. This enables membership cards to be issued, which engenders a beneficial feeling of belonging. When money has been given, then a basic list of equipment to start a club might include:

(1) Some of the commercial games mentioned in the previous section.
(2) Home-made games, using card, dice, counters, blank playing cards, centimetre cubes, transparent covering material for the card.
(3) Practical activities requiring card, string, cotton in different colours, needles, straws, pipe cleaners, spirograph, rulers, pencils, felt-tipped pens, crayons, glue, protractors, scissors, compasses, wooden board, nails, matchsticks and lots of paper (plain, gummed, graph, isometric).

If the club meets weekly, a competition may be set which requires pupils to take the problem sheet away with them and do some further work or research in order to find the solution. If club funds are available, then prizes will increase the popularity of the competition although one has to be careful to ensure that the real winner is not a parent!

8.5 Conclusion

It is not being suggested that a game is the only way to teach a topic in mathematics, but it is a weapon in the teacher's armoury which is too often ignored and which, if used in a well-organised manner, can make a real contribution to pupils' understanding and enjoyment of the subject. In order to increase their impact, games should be used selectively in the teaching of those topics where they can be of greatest benefit. The pupils will become absorbed in the playing of the game and will carry out the necessary mathematical operations with an entirely different attitude than would be evident if the same questions were listed in a text-book. During the course of a game pupils will often demonstrate a much higher level of mathematical ability than one had previously thought them to be capable of attaining. This, then, is the answer to the critic of mathematical games who says that to play these in the classroom is a waste of time. If motivation is the key to success in mathematics then the use of games is to be strongly recommended.

As with all teaching methods, however, success does not come automatically. The games and recreational work-cards suggested in this chapter must be attractively and neatly produced and covered with a protective film. When used in a lesson with a whole class, the equipment must either be put out before the children arrive, or given out quickly and efficiently at the beginning, as the orderly conduct of a lesson can rapidly be destroyed by the excitement engendered at the prospect of playing a game. This can be particularly difficult if the pupils are new to the game since progress cannot be made until all understand the rules. Where the game or puzzle is being given to an individual pupil, or to a small group of pupils, this situation is eased. The scope for stimulating pupils' interest in the subject is endless and this interest will often extend into their own time.

The pages of games or puzzles that appear in series of text-books are useful but have little to indicate how they might be used as a coherent part of the course. Although the examples in this chapter have largely been confined to games and recreational work that is simple and cheap to produce, easy to describe and fairly easy to play, an attempt has been made to show that games can be more than an incidental end-of-term activity and that there is much to be gained by including them in the mainstream mathematics course.

Bibliography

Ball, W.W.R. and Coxeter, H.S.M. (1974) *Mathematical Recreations and Essays*, University of Toronto, Toronto.

Banwell, C.S., Saunders, K.D. and Tahta, D. (1972) *Starting Points*, Oxford University Press, Oxford and New York.

Cundy, H.M. and Rollett, A.P. (1961) *Mathematical Models*, Oxford University Press, Oxford and New York.

Gardner, Martin (1959) *Mathematical Puzzles and Diversions*, Penguin, London.

Gardner, Martin (1961) *More Mathematical Puzzles and Diversions*, Penguin, London.

Gettings, F. (1963) *The Golden Pleasure Book of Art*, Paul Hamlyn, London.

Kordemsky, B.A. (1975) *Moscow Puzzles*, Penguin, London.

Lavin, M. (1972) *The Flagellation*, Allen Lane, The Penguin Press, London.

Lockwood, E.H. (1961) *Book of Curves*, Cambridge University Press, Cambridge and New York.

Pearcy, J.F.F. and Lewis, K. (1966) *Experiments in Mathematics*, Longmans,

Harlow and New York.
SMP (1970) *School Mathematics Project, Book A*, Cambridge University Press, Cambridge and New York.
Todd, A. (1968) *The Maths Club*, Hamish Hamilton, London.
Wenninger, M.J. (1971) *Polyhedron Models*, Cambridge University Press, Cambridge and New York.

9 LINKS BETWEEN MATHEMATICS AND OTHER SUBJECTS

Keith Selkirk

9.1 Introduction

The teaching of mathematics in schools may be justified in a number of ways. It is, for example, part of the cultural background of our civilisation, and as such should rank with art, music, literature and similar aspects of our heritage. Again it is a logical and efficient system of deduction and this may well transfer to problems outside the immediate area of the subject. The justification which appeals particularly to those whose primary interests lie outside the subject is, however, that it is useful. At a time when the limitations of our national and global resources are only too painfully apparent, this usefulness must be a major justification for the teaching of the subject in schools and for its important share in the total school curriculum.

The usefulness of mathematics is apparent not only in life after school, but also in its applicability to other areas of the curriculum. It is fashionable in some quarters to point out that many people make little use of mathematics in their everyday lives, and that much of what they do use is rule of thumb calculations of a type which receive scant attention in the classroom. Many jobs, however, require some mathematical ability. For example, Clarke and Toye (1981) found in a small local survey that only two out of 31 jobs investigated had no numeracy requirement whatsoever. Outside the field of employment it is just as easy to dismiss the relevance of mathematics. Not only is an understanding of elementary arithmetic of continual value, but an appreciation of spatial pattern is necessary in many situations. For example, when cloth is cut out to make a dress pattern, the folded edge of the cloth must be distinguished from the selvedges and an appreciation of the reflective symmetry of the pieces of cloth cut out is required. This soon becomes apparent when alterations to the pattern are needed, or the cloth is of a different width from that recommended.

It is not, however, the purpose of this chapter to dwell on the uses of mathematics out of school, but to concentrate on its position as a service subject within the school curriculum. It is important to remember that both aspects of usefulness have a valuable motivational force for the pupil studying mathematics. The writer remembers encouraging a fifteen-year-old intending architect to persevere in studying matrices by pointing out that they have been used by architects to determine the most efficient allocation of rooms within a building (March and Steadman, 1974), and teachers should be encouraged to make themselves familiar with such applications of their subject.

9.2 Recent Developments Linking Mathematics and Other Subjects

Twenty years ago, although mathematics was generally regarded as useful, this was acknowledged as far as other areas of the curriculum were concerned only in the case of physics. There the connection was so close that applied mathematics in school sixth-forms in Britain had come to be regarded as synonymous with that branch of physics known as mechanics. As a result a whole A-level subject was devoted by some specialist mathematicians in the sixth form to what is essentially a single mathematical model of the physical world. In some schools this is still the case, and it is certainly a useful study for prospective physicists and engineers, but it has led to a restricted view of the real nature of applied mathematics, and has been a disadvantage to those whose chief interests lay in other areas of the curriculum.

Today teachers are far more aware of the wide use of mathematics in these areas. The introduction in the late 1950s of A-level examinations in mathematics-with-statistics has led to the opening up of a whole new area of mathematics. It is an area of considerable pedagogic difficulty and we are far from solving all the teaching problems which statistics raises. A good start has been made at the O and CSE levels by the work of the Project on Statistical Education under the direction of Peter Holmes (Project on Statistical Education, 1980). More recently, the development of the computer, the micro-computer and the calculator has further emphasised the applicability of mathematics and opened up another new area of study. Some of this is subject specific (see, for example, Shepherd, Cooper and Walker, 1980, in the area of geography), and developments in the 1980s are likely to be rapid.

Just as in mechanics, there is a danger in statistics and computing that an over-concentration on specific areas of mathematics with important applications will continue to lead to a neglect of the nature of applied mathematics as a study. Essentially the applied mathematician has to build a model of a real life situation which may then be manipulated so as to provide insights into that situation. This important point merits consideration at greater length later in the chapter.

While mathematics teachers have become much more aware that their subject is being used by other teachers, relatively little has been done to help them to come to terms with the problems which this raises. According to Griffiths and Howson (1974): 'Ignorance concerning what users of mathematics actually need, as distinct from what academic mathematicians think they may need, is beginning to be combated by the publication of certain surveys.' These remarks are intended to apply to universities, but much the same situation is true of schools. It is timely, therefore, to summarise the literature on mathematics and its links with other subjects which has been written for school teachers.

9.3 The Literature: the Mathematicians' Contribution

While there have been many popular expositions of the relevance of mathematics to the outside world, most of the authors have not attempted to write in terms of the school curriculum, but have examined the applications of mathematics from a wider standpoint. Holt and Marjoram (1973) is one of the exceptions to this. There are, indeed, grave difficulties facing the author attempting to write about the applications of school mathematics, for no single author can readily comprehend the whole field of learning even at the school level. It is not surprising, therefore, that the two books which have been written in Britain expressly to cover the 11-16 mathematics curriculum are the result of Schools Council projects, and their authors had the backing of a project team to supply examples and advice.

The earlier of these books is *Crossing Subject Boundaries* (Mathematics for the Majority, 1974) which was written for the Mathematics for the Majority Project, the chief author being J.H.D. Parker. This project covers work done by average and below-average pupils in the 11-16 range. The book contains a valuable collection of ideas covering not only such obvious areas as science and geography, but also art, music, handicrafts, sport and hobbies. In these latter areas, where references are not easy to come by, it is particularly useful, but in science it tends to emphasise biology and environmental science at the expense of physics and chemistry, partly because it was expected that pupils of this ability would be less likely to study these areas. In fact the trend towards integrated science courses in the last decade for pupils aged 11–14 has meant a greater emphasis recently on the study of the physical sciences with such pupils.

The second book is *Mathematics across the Curriculum* (Ling, 1977) which was written for the project called 'The Mathematics Curriculum: a Critical Review'. This emphasises to a much greater extent the requirements of more able children, and consequently shows an awareness that for many pupils in this age range, mathematics must supply a preparation for the needs of the sixth form. The result is that these two books tend to complement one another. This latter book shows a much greater awareness of the role of mathematics in science, and over half of it is devoted to this area, giving a survey of the ways in which mathematics is used in the individual sciences and some examples of examination questions in O-level and CSE which use mathematics.

At a more basic level, the School Mathematics Project (1977) has also produced a short booklet *Modern Mathematics in the Science Lesson* which has much useful advice for teachers on the *minutiae* of their work, and pinpoints some areas of common concern to mathematics and science teachers.

At the sixth-form level (16–18 years) two other projects are worthy of mention. They have taken very different approaches to the problem of teaching applied mathematics. The Sixth Form Mathematics Project has published over a dozen booklets under the general title 'Mathematics Applicable' (1975–7) which were written for pupils of between O and A-level standards. Their major innovation has been their emphasis on applied mathematics as a model building activity, though these models have usually been drawn from real life situations and not specifically from other subjects of the school curriculum.

The Continuing Mathematics Project has published a flexible collection of 'Self-learning Units' in the form of booklets. These are written for students in the sixth form who require some additional work in mathematics to pursue their own subject interests, especially in biology, economics and geography. The materials cover specific mathematical topics and are written in a straightforward manner; they are very differently conceived from any of the materials mentioned above.

9.4 The Literature: Contributions from Outside Mathematics

In 1973–4, four reports were produced (Royal Society, 1973, for physics; British Committee on Chemical Education, 1974; Royal Society, 1974, for biology; Lewis, 1973, for social sciences). These attempted to summarise the mathematical needs of advanced-level students in the various subjects. The physics report urged closer collaboration between mathematicians, physicists and engineers in formulating the needs of A-level physics, stressing particularly the role of the GCE examining boards in this area. An extensive list of mathematical needs and desirable abilities was appended covering much of the traditional single subject A-level syllabus in mathematics, but making no mention of more recent introductions to the syllabus such as probability, linear algebra and numerical methods which have equally important applications in physics and engineering.

The chemistry report is more descriptive and raises important questions of wider applicability of the mathematics, it is well worth reading by those whose interests lie in fields other than chemistry. It emphasises the need for chemists and mathematicians to talk, co-operate and collaborate, particularly in learning to understand each other's language, it stresses the need for examination boards and text-book writers in chemistry to make their mathematical needs explicit, and it urges the need for in-service courses.

The biologists considered the whole field from O-level to tertiary education. In particular, they differed with the physicists in welcoming the new syllabuses in mathematics. They agreed with the chemists on the need for greater communication, and in addition pointed out the additional problems of getting pupils to realise the need for mathematical techniques and of biology teachers' reluctance to use mathematics. Like the social scientists they were also concerned with problems of syllabus content and examining too lengthy to summarise here.

More weighty and more specific are the many books which provide mathematics texts for students of other individual subjects. Most of these are written for students over 18 and are thus beyond the scope of this chapter. They are often mathematically uninspiring and traditional in outlook; sometimes they contain more or less obvious mathematical errors, and often they have only limited relevance to students they purport to help.

For teachers, Nuffield Advanced Science (1973) has produced *Supplementary Mathematics, Physics Teachers Guide* which is valuable particularly for those physics teachers wanting to teach additional mathematics to students not studying the latter subject at A-level.

A number of books take a more neutral stance between subjects, examples at levels suitable for younger children are Gibbons and Blofield (1971) in biology and Dienes (1973) in games, dance and art. Corresponding examples for teachers of older children are Dudley (1977) in biology and Selkirk (1982) in geography. Such books as these provide information and examples for mathematics teachers trying to present their subject as a living part of the total school curriculum. Unfortunately, the task of digging out the relevant materials is often a laborious one.

The Mathematical Association (1980) provides an extended reference list to books in this area which is kept on computer file so that it can be regularly updated. The busy mathematics teacher can, however, hardly be expected to read more than a few books relating his subject to other individual parts of the curriculum, and such books will inevitably tend to be read more by teachers of the other subject for which the book is intended. It does not seem unreasonable to expect every subject teacher to have some knowledge of the relationship of his or her own subject with mathematics in the same way as all teachers are expected to know something of the relationship of their own subject with the written and spoken language.

9.5 The Way Ahead: Research and Development

It is now time to outline the steps which might be taken to improve the links between mathematics and other subjects. Such steps fall into two categories; the first is steps which need to be taken outside individual schools and fall under the category of curriculum research and development, and the second is steps which need to be taken inside individual schools. In what follows these are considered in turn with appropriate examples, written from the mathematician's point of view; this is not an appropriate place to mention other than in passing the action needed by teachers of other subjects.

Two specific tasks of research and development need to be undertaken by projects since the amount of work involved in each is too complex for a single writer, and too time-consuming to be undertaken by full-time teachers.

The first has been partly accomplished by the projects which produced Mathematics for the Majority (1974) and Ling (1977) which

Figure 9.1: Assimilating New Practical Examples into School Mathematics

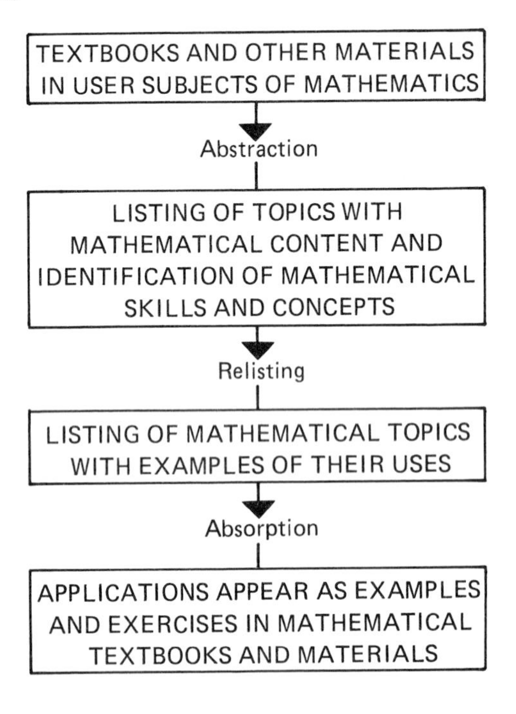

are described above, and the whole is outlined in Figure 9.1. Essentially such a project would examine in detail the various subjects of the curriculum. Its first stage would be to abstract from each subject's text-books and materials the mathematical content and examples of its use. The books cited above have already done much of this work up to O-level. Ling has, in addition, covered part of the next stage, which is to identify the mathematical skills and concepts which are required in each example. The third stage is to turn the lists about and under each mathematical topic list the appropriate examples of its use. At this stage the project might publish its work, but this would not be its final objective; that would be for the various applications to be assimilated into text-books and other source materials and be used regularly by the teacher in the classroom.

Both the books cited have many intriguing examples of uses of mathematics, opening the first at random gives as examples family trees, a graph showing the weight and age of a mouse and a discussion of the rigidity of various designs of stool. Here is plenty of highly motivational material. Why then have teachers made so little use of it? To ask this question is not to denigrate the authors or the teams who helped to provide their material. They had neither the time nor the resources necessary to complete their tasks, they were not able to go far enough to be of practical value to teachers in the classroom who face many pressures on the time they spend in the preparation of teaching material. For example, in Ling there are eleven different references to percentage, too many to be of value to a teacher planning lessons in that topic and looking for motivational examples. More useful would be a resource bank of such examples, carefully graded and readily accessible. This bank would be the objective of such a project, whose success would be judged not only by the use made of it by the classroom teacher, but also by the extent to which its work was plagiarised by other writers of texts and materials.

The second task is that of avoiding syllabus conflict between mathematics and other subjects, and this must be accomplished in parallel with the first task. The plan of action is illustrated in Figure 9.2. For each mathematical example encountered the questions must be asked (a) Has the relevant mathematics been taught? (b) If not, can it be taught? Scientists sometimes assume that the answer to the latter question is always yes, but of course this is not so. A well-known counterexample is the notorious oil-drop experiment which found its way into the first-year syllabus of the Nuffield Physics Course and which not only required volume formulae which pupils had not met, but also seemed to demand hard concepts of volume equivalence which most eleven-year-old children were not ready for. Less spectacular examples are still common in the classroom; it frequently happens that scientists expect an understanding of decimals, ratio or percentage beyond the conceptual abilities of their pupils. In such cases there is no alternative but to postpone the relevant work in the other subject, or to alter radically the teaching of it to avoid using the troublesome mathematics.

In other cases there must be negotiation. This can and should be done at school level, but it is frequently ignored, and it seems ridiculous that every individual school should have to make and correct the same mistakes when clear general recommendations would be a great help to all. As teachers of a service subject, mathematicians must be prepared to reorder their syllabuses where this is possible. There is also much work to be done so that mathematicians are more aware of the ways in which their subject is being used and both

Figure 9.2: Avoiding Syllabus Conflict

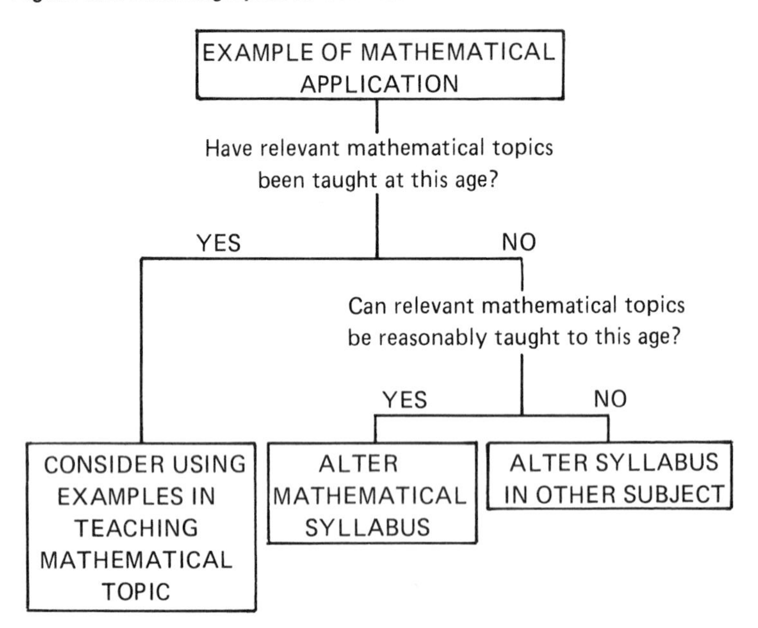

mathematicians and teachers of other subjects can adopt a common approach and language. In this way, pupils' learning difficulties may be diminished.

It will sometimes be necessary to leave the teaching of a mathematical topic to teachers of the other subject, but this is probably unwise if it can be avoided; mathematics teachers usually strive to teach understanding of the concepts involved, while users of mathematics, who have no vested interest in the subject, are tempted to adopt rule-of-thumb procedures relevant only to particular situations. This approach is often used in the sixth form with pupils who have no mathematics lessons (for example, with pupils studying physics but not mathematics at A-level), or because one subject option requires a particular piece of mathematics not needed by those studying other options (as in the case of the chi-squared test with biologists).

Examples

Before turning to action within schools, it might be helpful to look at examples of how these ideas might be worked out in practice.

The first example is taken from graph theory, an area of

mathematics which is becoming more popular in schools. There are many applications of the theory, one traditional one is that of Kirchhoff's laws in A-level physics; a newer one is that of networks in geography (which provides elementary examples of the use of matrices). It is a fertile field for examples of open-ended investigational work, providing many situations involving counting, generalisation, simplification and the development of simple rules and theorems. Cooke and Anderson (1978) give a simple introduction.

One of its most familiar problems, which has been used both in junior and secondary schools, is the pentomino problem. Five squares in a plane must be fitted together edgewise to make a connected shape. This can be done in twelve distinct ways. Teachers generally set up the investigation by showing the idea to pupils and asking them to find all the shapes, sometimes cutting them out so that rotations and reflections may be seen to be the same shape. Sometimes they set the difficult (but soluble) problem of fitting all the shapes into a 6×10 rectangle; sometimes they extend the idea to hexominoes, where there are too many shapes for most pupils to discover them all.

Generally the problem is left at this abstract stage, although sometimes pupils are asked which shapes may be folded into an open cubical box. The problem at first appears to have no other practical application, so it comes as a surprise to find the pentominoes in a university geographical text (Haggett and Chorley, 1969, p. 55) in relation to farm holdings in Missouri which have been combined. The equivalent problem with the hexominoes has been related to cell patterns in biology (Thompson, 1942) and there is some evidence that different combinations have greater or less stability.

Suppose now each square is replaced by a dot and adjacent dots are joined by straight lines. This is the same problem in a different form, and the pentominoes become shapes as in Figure 9.3. The rods can now be made freely jointed, with the joints at the dots. Figure 9.3 (f) is exceptional in that it has a closed circuit and so is ignored. Shapes (a) and (e) are equivalent, as are (b) and (d). Further investigation of the other six possible shapes will reveal no other configurations. The three possibilities are represented by Figure 9.3 (a), (b) and (c).

If the dots are now regarded as carbon atoms and hydrogen atoms are added as in Figure 9.4 to make the number of bonds around each carbon atom up to four, then molecular diagrams are obtained for three molecules all with the chemical formula C_5H_{12}. These are pentane and its two isomers which are structurally different. The actual molecules, although in reality three-dimensional, are often represented

Figure 9.3: Illustration of the Pentomino Problem using Dots and Rods

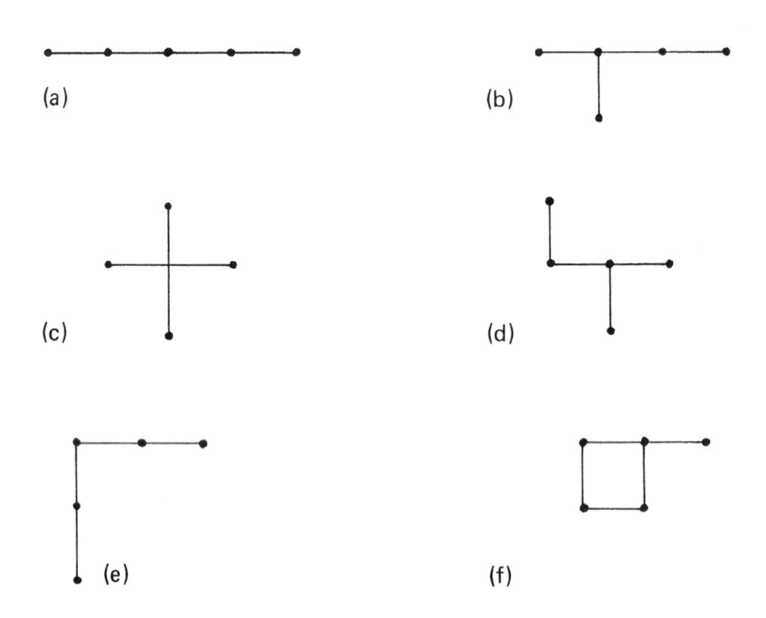

Figure 9.4: Pentane (C_5H_{12}) and its Two Isomers

```
      H   H   H   H   H
      |   |   |   |   |
  H − C − C − C − C − C − H
      |   |   |   |   |
      H   H   H   H   H
```

(a) pentane

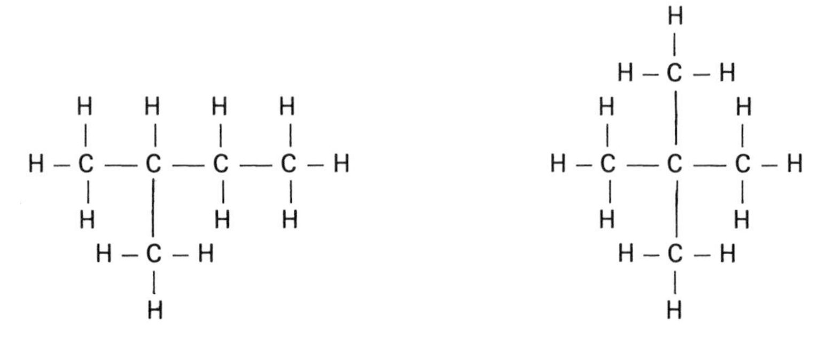

(b) 2-methyl butane

(c) 2,2-dimethyl propane

conventionally in this way. Such isomers, as well as being of some limited interest in sixth-form chemistry, provide a number of interesting investigations in mathematics.

In case this first example is rejected as being too novel, consider ratio and proportion as a more traditional area of difficulty. The problems of teaching this are summarised in Hart (1981) and from an applications point of view, in Ling (1977). Many applications of this topic will not be accessible to secondary children of below average ability, but the concepts are clearly of immense practical importance. Hart's examples include an unexciting one about buying calico and one derived from Piaget about feeding eels which is very contrived. Other examples are more realistic and it is clear that even in the sort of testing possible in Hart's project, ratio has many practical embodiments. In science, the repeated references in Ling testify to the importance of such calculations. And yet scientists frequently find that pupils have great difficulty with these calculations in the classroom, and sometimes they use poor results as a stick with which to beat the unhappy mathematicians.

In solving ratio problems children often resort to *ad hoc* procedures, notably halving and doubling, and science teachers tend to reinforce such methods. Children find halving and doubling much easier than other proportion problems, yet even Nuffield Advanced Science (1973, p. 17) reinforces this approach, thus laying up problems in other situations.

Chemists in particular make a great deal of use of ratio and proportion; Ling gives no less than six examples with solutions and comments (pp. 52-5). On p. 75 he also gives a biological example from a Joint Matriculation Board O-level paper (November 1973):

> A man of 25 years weighing 68 kg with a surface area of 1.8 m^2 has a surface area to weight ratio of 0.027 m^2/kg.
> A baby of one year weighing 10 kg has a surface area of 0.47 m^2.
> (a) What is the difference between the surface area to weight ratio of the baby and the man?
> (b) Give ONE important biological consequence of this difference.

Apart from the fact that $1.8 \div 68$ is 0.026 and not 0.027, the use of the word difference in this question is also unfortunate. Literally the difference is $(0.47 \div 10) - 0.026 = 0.021$ m^2/kg but this is hardly what the examiner wanted to know. Perhaps the examiner was almost as confused as pupils often are over ideas of ratio. The intention of the question is, however, entirely laudable, even down to the inclusion of a redundant number which is not needed in the solution. This is precisely the sort of applied mathematics which needs to be far more widely developed. Even the arithmetical calculations are realistic, and with the use of a calculator the difficulties of division of decimals disappear. The whole subject of length, surface area and volume or weight ratios is of considerable interest to biologists; see for example Thompson (1942) and Dudley (1977).

9.4 The Way Ahead: Steps in Schools

While research and development work can be of great help to school teachers, it is in the schools themselves that changes must be made if mathematics is to be effectively integrated into the school curriculum. In a series of four case studies by the Mathematics and Science Sub-Committee of the Mathematical Association some possible steps for action were identified. Details appeared in two articles in 'Mathematics in School' (1982). I am grateful for permission to summarise the main points for the four schools identified as A–D:

School A: A member of staff is responsible for liaison, in this case the head of science. An exchange of syllabuses is effected, an informal discussion develops about mutual problems. Texts such as Ling (1977), Mathematics for the Majority (1974) and School Mathematics Project (1977) are available.
School B: Regular meetings take place once a term between the heads of departments involved. A checklist is instituted by the head of science making inquiries about the timing of the teaching in his department requiring certain mathematical skills.
School C: Science and mathematics are in the same faculty, which facilitates involvement. Joint materials are written for one or two periods per week for the first-year pupils; this involves meetings, discussions and firm leadership. The chief theme is measurement. At termly faculty meetings, each major group supplies a question for discussion involving a mathematical concept which is found difficult (e.g. ratio).
School D: An integrated course was planned for the first year, in this case twelve year olds. By block timetabling staff were made available together, and work was divided between mathematics, science and integrated studies. Booklets for the latter area were

produced by pairs of staff, one scientist and one mathematician, who initially team-taught the relevant classes.

These four schools illustrate a progressively developing liaison between mathematics and, in this case, science. Two important points appear again and again in studies of the problem. The first is that of persuading members of staff to talk together about their mutual teaching problems, and the second is the necessity for clear and effective leadership in any collaborative exercise. It is surprising how difficult talking can be. Scientists make unthinking criticisms of mathematics teachers about the failure of their pupils to understand problems with a mathematical content, while mathematics teachers often have little time for those who use their subject, and little interest in the uses. The writer has been accused on separate occasions, apparently seriously, of making take-over bids for chemistry and geography, as if mathematics teachers did not have sufficient problems of their own! If talk can generate heat, how much more can leadership, but this is another task to be undertaken by over-burdened and often criticised heads of department.

In the above, links with science have been mentioned, but there needs to be similar collaboration on a smaller scale with other curriculum areas. For example, Chapter 1 of the Oxford Geography Project (1974) could form the basis for a first half-term of teaching in mathematics and geography for new first-year pupils at a secondary school. Many of the ways in which this chapter looks at the environment in and around the school could be used to motivate parts of the mathematics syllabus. Practical subjects also offer great opportunities when teachers are aware of them; cookery, needlework, metalwork, art and music all have practical mathematical content. The writer vividly remembers a student teacher using the positions of frets on a guitar as an introduction to the theory of logarithms.

9.5 Problems of Language and Notation

If talk between teachers is important, what should they talk about? This section outlines some of the problems which mathematics teachers can discuss with their colleagues from other disciplines. Since talk is the order of the day, matters of language are perhaps top of the agenda for such discussion.

Both mathematics and other subjects have their own jargons. Words are used with precise and sometimes differing meanings. More often different words are used with the same meaning. All this is very confusing for pupils. One example is the use of the word difference mentioned above, but there are many more. Science teachers, who have often been reared on traditional geometry, will tend to use unfamiliar words and phrases such as corresponding angles and congruent triangles, which pupils following a modern course will not have met. A list is given in School Mathematics Project (1977). Paradoxically it is now quite likely that in schools still teaching traditional mathematics syllabuses the younger science teachers will have been reared on modern syllabuses and the procedure will have to be reversed.

Even apparently trivial problems of notation can cause difficulty when teachers are not aware of them. Some common examples are (a) the use of $^{+}4$, $^{-}7$ for directed numbers instead of the traditional $(+4)$, (-7); (b) the description of such numbers as positive four and negative seven rather than plus four and minus seven (both these so that the operations of add and subtract can be seen as different from the signs of the numbers); (c) the use of different letters to represent the same thing, for example do we say $P = mf$ or $F = ma$?; (d) the x-axis in a graph might be called $y = 0$ or Ox, and pupils often get confused between these; (e) points in graphs may be marked in several different ways: • × ⊙ +. Mathematicians generally prefer ×, but even they are not consistent. Sometimes the alternative markings are useful to distinguish between different graphs in the same diagram; (f) do we say two times x plus y, two bracket x plus y or two into x plus y for $2(x + y)$. Does the last form imply division by the use of 'into'? (Confusion about the meaning of difference has already been pointed out.); (g) cross-multiplication is often a point of disagreement; scientists are often fond of this process without always being very clear about its limitations; (h) mediator is replacing perpendicular bisector; (i) dilation and enlargement are alternatives; (j) the words used to describe networks are legion; vertices, nodes, points, junctions and intersections; edges, links, arcs, routes and branches; faces, regions and areas. The reader will no doubt be able to think of further examples. Teachers need to be very sensitive to blank looks when they are using technical language across subject boundaries.

9.6 Arithmetical Techniques

Perhaps the major criticism of mathematics teaching by users of mathematics is that pupils are weak in arithmetic, or as it is often put, 'they don't know their tables'. There is, in fact, much evidence that a majority of children will never be able to understand and carry out complicated algorithms such as long division. *Aspects of Secondary Education in England* (Department of Education and Science, 1979) says 'whatever provision may be made in special schools there will always be some children in ordinary schools who achieve only minimally in the subject'. It looks as if weak arithmeticians, like the poor, are always with us. As is pointed out elsewhere in this volume, however, the advent of the calculator has brought a new factor into school arithmetic, and both mathematicians and other teachers must come to terms with it. Not only does it solve the difficulty of the complicated algorithms, but it makes real-world calculations more possible. Its use will clearly figure largely in any inter-disciplinary discussions between teachers, and Chapter 7 might be used as a basis for these. A first objective might be an agreed school policy on the use of calculators. At the very least the scientists must be made aware that logarithms are no longer taught in some O-level courses.

The advent of the calculator does not signal the demise of arithmetic. Both mathematics teachers and others must become much more aware of how children use number operations and the mistakes they make. As Margaret Brown says (in Hart, 1981): 'In a situation where calculators are readily available for computation, it is clear that the emphasis must change from algorithm learning to understanding the structure of the operations themselves and how and when they should be applied.' She makes it clear that up to half the children entering secondary school are shaky on concepts of multiplication and division, even with whole numbers, while the worst 10 per cent have little appreciation at all of these concepts.

For average and above average children there is a need both in mathematics and its applications for a far greater emphasis on accuracy and estimation. Scientists are often unaware of the magnitude of rounding-off errors in calculations. For example

$$\frac{72.3 \times 16.8}{13.2} = 92.0$$

is a likely calculation to occur in an experiment. The data and solution are each given to three significant figures. On mathematical grounds alone, however, the limits for the solution are

$$\frac{72.25 \times 16.75}{13.25} = 91.33$$

and
$$\frac{72.35 \times 16.85}{13.15} = 92.71$$

Not only is the third figure in the original solution unjustified, but even the second one might not be correct (though admittedly the probability of this is low). Pupils, and even teachers, can forget this and blame errors in experiments on poor experimental technique or unsatisfactory equipment. Calculators make a naive approach to such problems quite possible for many pupils.

The dangers of a blind use of machines means that estimation of solutions is even more important than it was. The calculation above might be crudely estimated by

$$\frac{70 \times 20}{10} = 140$$

or by a more able pupil, with rather greater accuracy as

$$\frac{72 \times 16}{12} = 96$$

but both show that the correct answer is roughly of the same magnitude.

These points are worth extended discussion, but above all, teachers of other subjects must learn to make realistic mathematical demands of pupils, while mathematics teachers must realise that all arithmetical calculations used by teachers of other subjects are applied mathematics whose basic teaching must be done by themselves.

9.7 Statistics

Not very long ago the use of statistics in schools was virtually confined to averages, by which was meant arithmetic means, and to a few simple statistical graphs. Nowadays these and other simple statistical ideas are widely used in project work in all manner of subjects (for example, see Project on Statistical Education, 1980). This wide range of applicability makes the problem of achieving consensus in teaching harder, though fortunately the mathematical depth is in most cases not too great.

Statistics, however, harbours traps for the unwary, and mathematics

Figure 9.5: Incorrect and Correct Ways to Illustrate Quantities in the Ratio 1 : 2 : 3 by Means of Solid Diagrams

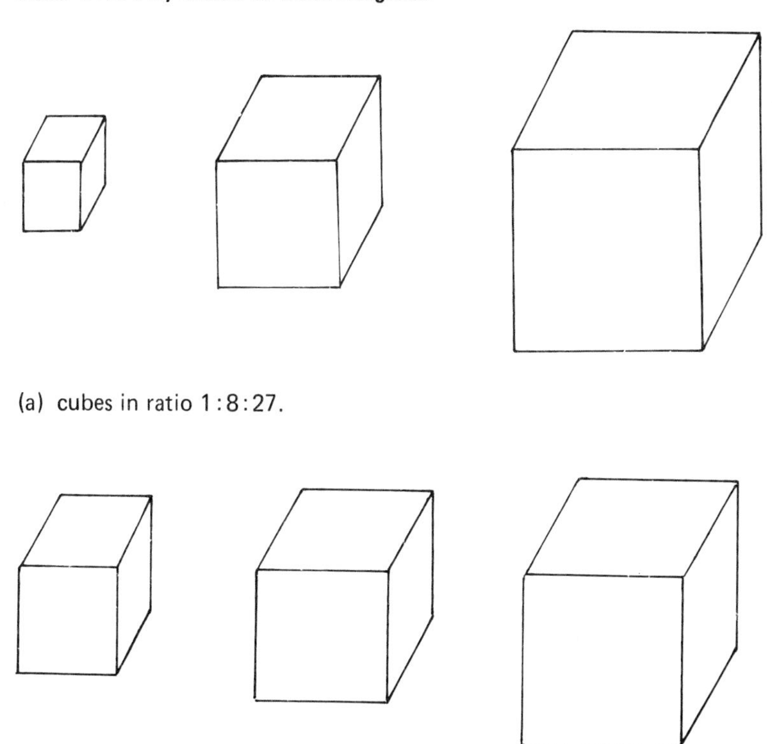

(a) cubes in ratio 1:8:27.

(b) cubes in ratio 1:2:3.

teachers have a duty to make their colleagues aware of these. Figure 9.5 illustrates one well-known error, which is related to the surface area to weight ratio problem quoted above. Such diagrams are anyway best avoided because of the tendency of the eye to compare the two-dimensional areal representations rather than the three-dimensional realities. It is easy, too, to get bogged down in minor details such as how to find the median of a grouped distribution, or where to place the end points in a cumulative frequency graph. Such trivia cause difficulties for text-book writers, let alone inexperienced teachers.

An area of valuable expertise which should receive more attention, but is often forgotten, is the scanning of tabulated data to extract the important points quickly. This skill is of particular importance in the study of economics. The illustration of data has many advantages in improved understanding, but is often poorly taught. Statistical diagrams and graphs should be clear and unambiguous, and convey a visual message which gets the point over more easily than the corresponding numbers.

More advanced statistics, notably inferential statistics, is a sixth form matter, and special difficulties arise when other subjects place demands on mathematics in this area. Examples are the uses of correlation coefficients by geographers and of the chi-squared test by biologists. The mathematical justification of these is quite difficult, certainly that of the latter is post-school work. The materials of the Continuing Mathematics Project help here, but teaching at school level cannot attempt complete understanding of these ideas. Even accepting that a knowledge of statistical theory is not necessary to be able to use these techniques, one wonders how many sixth-form pupils have any depth of understanding of their application.

9.8 Algebra

Algebraic problems of notation and language have already been mentioned, but algebra itself is a major area to consider in the relationship between mathematics and subjects which use it. One might characterise much of science as the attempt to understand natural phenomena by the construction of algebraic models of them.

Equations play two different roles in science. Some equations are definitions such as

$$R = \frac{V}{I}$$

in electricity which defines resistance, or are immediate consequences of definitions such as the familiar equation connecting degrees Celsius and degrees Fahrenheit. Others are quite different in that they are merely attempts to model natural phenomena. An example from physics is

$$PV = kT$$

relating pressure, volume and absolute temperature in gases. Such equations are approximate, often wildly so, except under unattainably ideal conditions. Even the familiar equations of kinematics fall into this class. Mathematicians are not always as clear as they might be about these differing types of equation.

Scientists are often unaware that in many courses, pupils solving equations begin with the flow diagram method. Thus $3x + 7 = 19$ might be solved

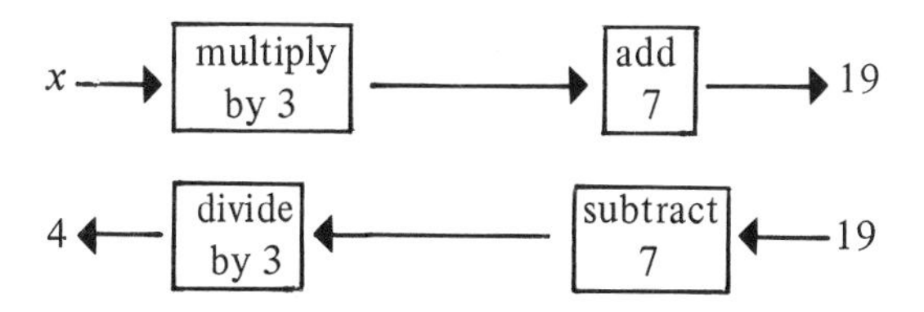

It is now easy to see that the answer is 4, but teachers only aware of traditional methods of solution can confuse pupils who have met this method alone. It should also be pointed out that this method of solution has severe limitations when equations become more complicated.

It is often not realised how difficult most pupils find the manipulation of algebraic equations, even in the sixth form. Even in mechanics pupils prefer numerical examples, and are seemingly unaware of the dimensional checks which are possible in literal equations. Scientists often prefer to make the unknown quantity in a formula the subject of that formula before solving it numerically. Thus in the equation for resistance mentioned above, if $R = 4$ and $V = 12$, then the equation would be rearranged to read

$$I = \frac{V}{R}$$

before substituting to obtain $I = 3$. Most pupils would find it very much easier to substitute in the original equation and simply solve

$$4 = \frac{12}{I}$$

The first method is, in any case, only advantageous when repeated applications of the formula are required.

There is an important area of contention between mathematicians and scientists about whether a letter such as 'R' stands for 4 or for 4 ohms, or in general terms whether letters stand for numbers or for quantities with units attached. This is a disagreement which usually gives value for money in acrimonious arguments. There is no way out of the dilemma, except to be aware that there is a choice, and to be clear what choice has been made. If the letters stand for numbers, then the numerical calculation is separated from that of having correct units, and the compatibility of the units must be checked separately. If on the other hand the letters stand for quantities, for example 4 ohms and 12 volts, then the solution will be given in the compatible unit, in this case the amp.

Algebra is related to statistics through the study of graphs. While many subjects have used a wider variety of graphical representations such as arrow diagrams in recent years, physicists in particular tend to use only the functional graphs and may not be aware of others. Graphs of functions are among the most important applications of mathematics and are important as visual representations of algebraic models. Often they are used with doubtful accuracy, for example, the exponential model of population growth deriving from the work of Malthus is clearly ridiculous, implying as it does an infinite upper bound to population. Even so, it reminds us that there must be an upper limit to growth and that an S-shaped graph such as the cumulative normal distribution would be a better model.

One area of algebra which has entered the school syllabus in recent years is that of sets. These have caused problems to non-mathematicians, and it is not unusual to find even university texts with errors. While their usefulness is perhaps exaggerated, we as mathematicians must be grateful to those who have explored applications of this fundamental concept on our behalf. Dudley (1977) has some interesting examples of applications to biology of both sets and also in the area of mappings and functions.

9.9 Geometry

It is more difficult to categorise spatial ideas than numerical ones, and this sometimes leads to the neglect of geometry. As a study of a logical mathematical system, the days of Euclidean geometry in schools seem to be numbered. But all is not lost. For example, Pythagoras' theorem is indispensible as it provides the means of calculating distances in a rectangular Cartesian co-ordinate system, and the properties of many simple figures remain important.

More recent introductions into the syllabus have brought new insights into applications of mathematics, particularly into the fields of craft and design. Pupils are often more aware of the direct relevance of geometry to the world about us. One useful topic which at one stage seemed in danger of being lost is that of the simpler ruler and compass constructions of elementary geometry which depend on the properties of the isosceles triangle and the kite. These are useful in technical drawing, and at the very least aid psycho-motor skills, for which our scientific colleagues will be grateful.

Motion geometry has often changed our perspectives, corresponding angles may be thought of in terms of translations, and vertically opposite angles in terms of rotations. While its applications are sometimes contrived, we have not always made the most of those which are available to us, for example in the study of light. Matrices, which can link motion geometry to algebra have found few uses at school level despite their undoubted value in more advanced work. Only in their application to simple networks, themselves another branch of geometry, have they found a fringe application in geography.

Similarity remains an important idea with many applications, though nowadays it is thought of as a combination of isometries and enlargements and the word itself may not be used. The idea of scale factor is now important, with obvious applications for geographers in map work, but many teachers using applications of mathematics still tend to be unaware of the altered approach.

Scale factor also appears in trigonometry where the old 'sine equals opposite over hypotenuse' definitions are often abandoned for the scientifically much more revealing definition based on the ordinate of the unit radius vector which can be applied to angles outside the range 0° to 90°. Right-angled triangles are then developed by using a scale factor on the similar right-angled triangle with unit hypotenuse. Teachers of other subjects (or of mathematics too!) who try to revert to the original definition can cause considerable confusion.

9.10 Investigations and Modelling

Finally it is worth returning to an idea which has already been mentioned several times in this chapter. An example of an investigation is the pentomino work outlined above, and similar work now takes place in many schools, particularly for CSE projects. Unfortunately less has been done with more able pupils at GCE O-level. This type of work has been much encouraged by the Association of Teachers of Mathematics who publish several pamphlets of ideas.

This is a welcome addition to teaching methods, but one of the criticisms that can be made is that investigations often lack practical

Figure 9.6: Mathematical Modelling as a Problem-solving Activity

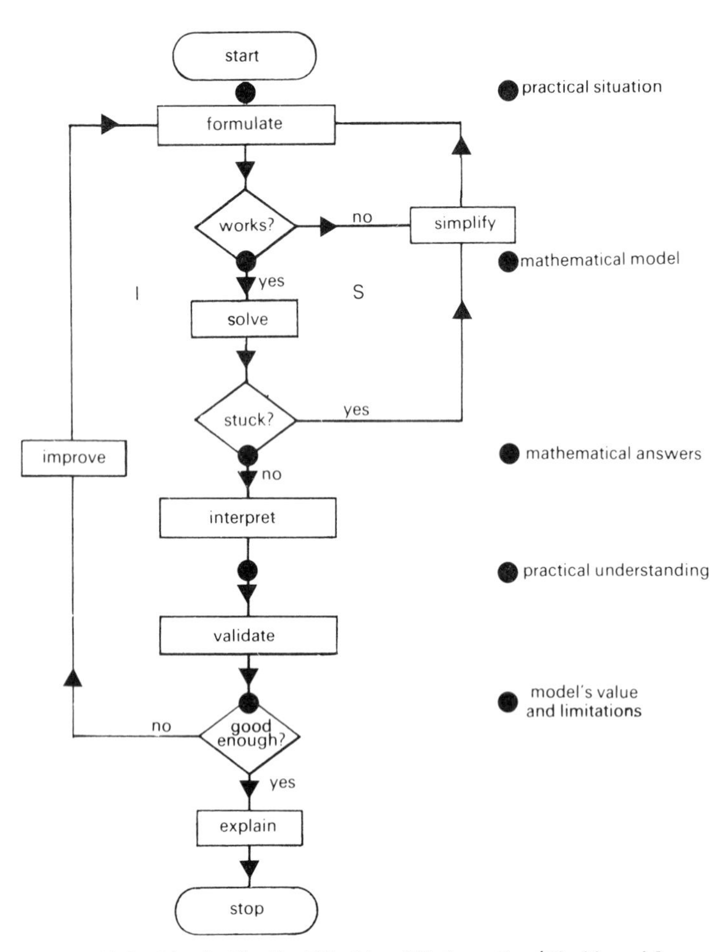

Source: H. Burkhardt, *The Real World and Mathematics*, (Blackie and Son, Glasgow, 1981).

content, and do not offer an improved understanding of how mathematics is applied in practical situations. There is a great need for development in this direction, especially for pupils in the 14–18 age range with above average and perhaps average ability. The Sixth Form Mathematics Project, which has published a series of books under the general title 'Mathematics Applicable' (1975–7) has been outstanding in this respect and has led to an increased interest in mathematical modelling in the sixth form. While some of their examples seem a little contrived, it is easy to be critical of a project which has developed a refreshingly new approach to applying mathematics. As far as other subjects in the curriculum are concerned, however, the necessity to avoid problems which involve too deep a non-mathematical content has been a drawback.

Burkhardt (1981) has attempted to codify modelling ideas into a theoretical framework, though in fact most of his book is concerned with practical examples (Figure 9.6). He emphasises that whereas pure mathematics is concerned with the *solution* of problems, modelling involves a number of other activities and a wider approach to the problem-solving process. Firstly there is the *formulation* of the mathematical problem which then, if possible, must be solved. It next has to be *interpreted* to see how it applies in real life, *validated* to assess its practical value and finally *explained*. In addition, if a solution of the pure mathematical problem proves impossible, it must be *simplified*, while if validation is poor an *improvement* of the model must be attempted.

Clearly the whole process is a much more complicated task than is usually realised, and requires more time than the traditional problem in pure mathematics. Because of these complications and the open-ended nature of the process, unfamiliar stresses are thrown on the mathematics teacher, who also has to cope with new and more complicated modes of assessment. Further, because of the number of steps involved, the solution stage becomes of diminished importance, and it is necessary to accept that the pure mathematical processes involved should be of a simpler nature than is usual for a given ability pupil, though to some extent this can be counterbalanced by allowing longer for the completion of an individual problem.

Many modelling activities demand the setting up of algebraic equations which have to be solved. Graphical models are often an alternative version of this type. Some models, however, are more geometrical in nature, though even then there is usually an underlying algebraic model. Geographers, in particular, use the word modelling frequently. Bradford and Kent (1977) detail some of these models from the geographer's standpoint, while Selkirk (1982) gives a more mathematical view of a wider variety of models.

An example of a model with geometric and algebraic content is that of city growth. The *formulation* of such a model might well begin by supposing that the crucial factor is the distance from the city

Figure 9.7: The Concyclic Model of City Development and Some of its Variations

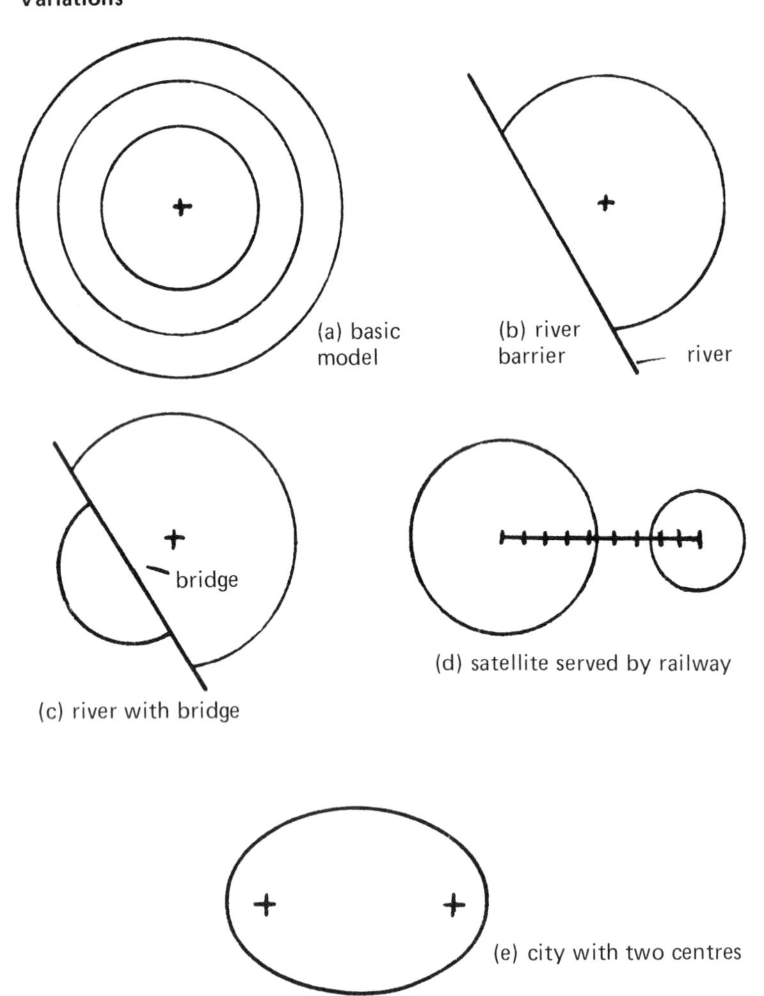

centre, leading to a geometrical *solution* of circular shaped cities, which can also be expressed in algebraic terms. Such a model hardly needs *simplification*, but can be *interpreted* by suggesting a growth in concentric rings and *validated* by specific examples such as Glasgow. The most interesting aspect of this very simple model is its potential for *improvement*. The effect of barriers to growth may be examined, for example rivers as at Liverpool and bridges as at

Worcester. Improvements in communications suggest that time of travel is almost as important as distance and examples cited of satellite towns around stations with good rail communications to major cities bear witness to this. The model may also be extended to take account of cities with more than one centre, for example Greater London may have been affected by the twin centres of the City of London and of Westminster. These variations on the model are illustrated in Figure 9.7.

9.11 Conclusion

There is much work which needs to be done on the relationship between mathematics and other subjects in the curriculum. Some of this work necessitates a co-ordinated effort, some demands detailed examination in individual schools. Perhaps too rigid a distinction has been made above between these, the boundaries everywhere are not clear-cut but blurred. The introduction of new ideas in mathematics teaching and in other subjects has proceeded independently, and mathematics is hardly nearer to being an effective service subject now than it was 25 years ago, however well it might be achieving its other aims. Mathematics must not be dominated by its service aspect, but it must take due account of this role if it is to give value for money as the second most important subject (after English) in the curriculum.

There is much work to be done in helping mathematics teachers to provide this service, both in the production of books and other teaching materials and in the provision of in-service education. As mathematics teachers we must not fail our colleagues in other subject areas by not meeting their needs to the best of our ability. In return we can expect an increased motivation for our pupils to study our own subject for its usefulness; in the fullness of time, some of these will be encouraged to love the subject and learn to study it for its own sake.

Bibliography

Bradford, M.G. and Kent, W.A. (1977) *Science in Geography 5: Human Geography: Theories and Their Applications*, Oxford University Press, London and New York.

British Committee on Chemical Education (1974) 'Mathematics and School Chemistry', *Bulletin of the Institute of Mathematics and its Applications*, *10*, No. 3, 80–6.

Burkhardt, Hugh (1981) *The Real World and Mathematics*, Blackie & Son, Glasgow.

Clarke, Aileen and Toye, Janet (1981) 'Jobs and Numeracy in the Classroom', *Mathematics in School*, *10*, No. 1, 10–11.

Continuing Mathematics Project (1977) A collection of self-learning units, Longmans Resources Unit, York.

Cooke, Charles and Anderson, Ian (1978) *Counting and Configurations*, Blackie & Son for the Schools Council, Glasgow.

Department of Education and Science (1979) *Aspects of Secondary Education in England: a Survey by HM Inspectors of Schools*, HMSO, London.

Dienes, Z.P. (1973) *Mathematics through the Senses, Games, Dance and Art*, National Foundation for Educational Research, Windsor.

Dudley, B.A.C. (1977) *Mathematical and Biological Interrelations*, John Wiley, Chichester and New York.

Gibbons, R.F. and Blofield, B.A. (1971) *Life Size: a Mathematical Approach to Biology*, Macmillan, Basingstoke.

Griffiths, J.B. and Howson, A.G. (1974) *Mathematics: Society and Curricula*, Cambridge University Press, London and New York.

Haggett, Peter and Chorley, R.J. (1969) *Network Analysis in Geography*, Edward Arnold, London.

Hart, K.M. (Ed) (1981) *Children's Understanding of Mathematics: 11-16*, John Murray, London.

Holt, Michael and Marjoram, D.T.E. (1973) *Mathematics in a Changing World*, Heinemann, London.

Lewis, J.P. (1973) 'Mathematics for Social Science Students', *Mathematical Gazette*, *LVII*, No. 401, 160–5.

Ling, John (1977) *Mathematics across the Curriculum*, Blackie & Sons for the Schools Council, Glasgow.

March, Lionel and Steadman, Philip (1974) *The Geometry of the Environment*, Methuen, London and New York.

Mathematical Association (1980, regular updating intended) *Booklists for the Teaching of Mathematics in Schools: Applications of Mathematics*, Mathematical Association, Leicester.

Mathematics for the Majority (1974) *Crossing Subject Boundaries*, Chatto & Windus for the Schools Council, St. Albans.

Nuffield Advanced Science (1973) *Supplementary Mathematics, Physics Teachers Guide*, Penguin for the Nuffield Foundation, Harmondsworth.

Oxford Geography Project (1974) *1. The Local Framework*, Oxford University Press, London and New York.

Project on Statistical Education (1980) *Teaching Statistics 11-16*, Foulsham Educational for the Schools Council, Slough.

Royal Society (Institute of Physics, Physics Education Committee) (1973) 'The Mathematical Needs of 'A' Level Physics Students', *Bulletin of the Institute of Mathematics and its Applications, 9*, No. 4, 109–12.

Royal Society (Institute of Biology, Biological Education Committee) (1974) *Report of the Working Party on Mathematics for Biologists*, Institute of Biology, London.

School Mathematics Project (1977) *Modern Mathematics in the Science Lesson*, Cambridge University Press, London and New York.

Selkirk, K.E. (1982) *Pattern and Place: an Introduction to the Mathematics of Geography*, Cambridge University Press, London and New York.

Shepherd, I.D.H., Cooper, Z.A. and Walker, D.R.F. (1980) *Computer Assisted Learning in Geography*, Council for Educational Technology with the Geographical Association, London.

Sixth Form Mathematics Project (1975–7) Mathematics Applicable Series of books. Heinemann Educational / Schools Council, London.
Thompson, D'Arcy, W. (1942) *On Growth and Form*, Cambridge University Press, London and New York.

10 THE ASSESSMENT AND EVALUATION OF SCHOOL MATHEMATICS

David Carter

Within a single chapter it is not possible to cover adequately all aspects of assessment even within the discipline of mathematics, nor to survey the literature other than selectively and briefly. No attempt has therefore been made to analyse the variety of assessment techniques that are available to the teacher nor to describe statistical methods commonly used in data analysis. Further reading in these and other areas is suggested at the end of this chapter. We shall be concerned with a discussion of the current issues in assessment of school mathematics in an attempt to make the classroom practitioner more aware of the need to find out a great deal more about pupils' achievement and difficulties, the effectiveness of his/her own teaching and the suitability or otherwise of the school mathematics curriculum.

10.1 Are Standards in School Mathematics Declining?

McIntosh (1979) reveals that the debate about standards of numeracy has been on-going for the past 150 years. Many of the quotations used by McIntosh might have been taken as contemporary were it not for some revealing terminology such as 'elementary schools'. Although McIntosh is writing about primary school mathematics much of the criticism quoted is close to that levelled at school leavers by employers and others, e.g. IMA (1976, 1977, 1978). It is clear that the problem has existed for a long time. Is it likely that there has been a steady decline in so-called standards of numeracy over this period of time? Is it possible that for a particular group of pupils the level of facility demanded is beyond their capability? It depends very much on what the term standards means; whether an acceptable level of performance or what it has now become possible to achieve. Sturgess (1978) suggests that for low ability pupils not to be 'socially deprived', the number of basic numerical skills required is very small. It has been known for some time that craft and technician students experience specific difficulty with calculations, express real anxiety about the subject, and often criticise the quality of the instruction they receive. Rees (1973) conducted research into this area of difficulty by constructing a diagnostic calculations test based on the relevant syllabuses. Sufficient precautions were taken to ensure that the test had reliability (repeated applications of the test would produce the same or very similar results) and validity (the test would measure what it is supposed to measure). This is not necessarily true of employers' tests which are either often homemade and therefore could be of doubtful reliability or are commercially produced and may be of doubtful validity unless the user is aware of the purpose for which the test was originally constructed. The results obtained by Rees showed that craft, technician and Ordinary National Certificate students performed least well, within each group, on a common core of items which included operations on numbers less than unity, reciprocals and the finding of areas. Moreover the teachers of these students tended to underestimate the difficulty of these items. Subsequent administrations of the test to groups of secondary-school pupils and first-year university engineering students produced remarkably similar results, i.e. relative to each group, performance was worst on the common core of items. Rees (1976) also administered the test to teachers training for the primary, secondary and further education stages again with very similar results. Some important questions are raised as a result of these particular studies. Are certain skills and concepts difficult to teach and to learn? Do teachers underestimate the difficulty their students have in acquiring certain skills and concepts? Much public criticism is by implication levelled at the teaching of mathematics in secondary schools. Rees's studies show that teachers both in secondary and primary schools have similar difficulties. It is generally recognised that the difficulties experienced and the attitudes acquired by pupils at the primary stage are major factors in subsequent success or failure. Hughes (1979) conducted similar research, but on a smaller scale, on undergraduates, pupils, teachers and groups outside the educational context. She found that tasks involving fractions, percentages and ratio caused the greatest difficulty for each group.

The process of assessment is inextricably linked with research findings. In spite of the fact that results of research are more properly discussed in Chapter 4, there is one single major research study the findings of which have serious implication for what and how we assess – that reported by Hart (1981). I have singled out this research because it is of outstanding importance and should be brought to the attention of all teachers – mathematics specialists or otherwise – in the primary, secondary and further education stages. The most significant general conclusion that Hart reaches is that '. . . mathematics is a *very difficult* subject for most children.' (p. 209). She further states that 'As teachers we have expectations of what a child "should" know, very often based on intuition and usually very different from the actuality.' So too with many critics of pupils' numerical abilities who tend to assume that the mathematical skills, concepts and understanding accessible to them are accessible to all pupils sooner or later.

The problem of assessing pupils' mathematical abilities may be one of comparing like with like. The aims of teaching mathematics have changed over the past 20 years with consequent changes in the curriculum. The demands of employment have also changed. These two developments may not have exactly proceeded hand-in-hand, thus possibly resulting in a mismatch. During this period, pupils of above average ability, who previously may have entered employment at 16+, responded to the call to enter the rapidly expanding further and higher education stages, thus creating more opportunity for the less well-qualified school leaver. This trend is likely to be reversed in the light of current financial restrictions. One trend that shows no sign of reversal is that over the past 15 or so years the number of pupils obtaining a GCE O-level pass or a CSE grade 5 or better has risen remarkably to somewhere around 80 per cent of the age group. This seen against the background of recent criticism seems to imply either that national certification of mathematical competence reveals very little about pupils' standard of numeracy or that much of the criticism is levelled against the well below-average ability school leaver.

There exist many local initiatives which seek to establish standards of performance at vital stages of transfer between the various stages of the educational system. The real problem is that no national base line of performance at any level – primary or secondary – has been established in the past against which to compare the performance of pupils in our classrooms today. In any event it is not enough to say merely that standards are falling or rising – if it indeed is the case. We need to know in which particular aspects standards are falling or rising; we also need to know whether these aspects are currently more or less important than hitherto. There is a real danger that the publicity given to the testing of basic skills may convey '. . . the message that they constitute the most important part of the curriculum, and so potentially distorting the overall curriculum balance.' (Kay, 1977, pp. 20–1).

10.2 Recent Surveys of Mathematical Performance

An account of the work of the first phase of the IEA (International Project for the Evaluation of Educational Achievement), on the teaching of mathematics, can be found in Husen (1967). This was the first study of its kind and was an attempt to evaluate pupils' performance across a diversity of national educational practices; England being one of twelve participating countries. The data on 12,000 English pupils aged 13+ and 18+ from 400 selective and non-selective schools in the independent and maintained sectors used in the international study together with additional data on 3,000 pupils aged 15+ to 16+ in 200 schools was used in a follow-up study (Pidgeon, 1967). The study involved assessment of performance in mathematics tests which were limited in content by the need to produce a test that satisfied mathematics teachers in twelve different countries. In addition pupils' opinions on their perception of (1) what teachers valued and emphasised and (2) school learning were assessed. Thirdly some attempt was made to evaluate pupil attitudes (1) towards mathematics as a process, (2) about the difficulties of learning mathematics and (3) towards the place of mathematics in society; in a more general context, (4) towards school and school learning and (5) towards man and his environment. The main conclusion drawn from this study provides a clear message for mathematics teachers in secondary schools in that it was found that the mathematics tests were too difficult for pupils in the 13+ age group; this was true for all countries. It appeared that pupils had less mastery of the subject than the test constructors expected them to have. It must also be noted that as a result of the pilot study some of the more difficult items were rejected. Thus perceptions of pupil attainment were even further from reality initially. Pupils of teachers who said they were teaching 'new maths' produced better performances in groups of items, whether 'new' or otherwise, than did pupils whose teachers said they did not teach 'new maths'. Subsequent analysis showed that teachers who said they taught 'new maths' tended to be specialists and also showed a greater interest in their subject. One must observe that this evidence ignores data from pupils whose teachers said they were teaching both 'new' and 'old' maths or who omitted to answer the question. Further, no clear definition of new maths was supplied to the respondent and therefore much depends on what the teacher's own perception of new maths was. Pidgeon claims that the mathematics test had high validity for English pupils since correlation of results with subsequent grades obtained in

GCE O- and A-level Pure Mathematics examinations was high. One does not doubt that as a result of this evidence the mathematics test was valid for those pupils – but can it be said to be necessarily valid for those pupils not subsequently entering GCE examinations?

Analysis of data on the five attitude scales showed that there was little inter-relationship, and further that there was little relationship between any single scale and scores on the mathematics test. One might expect that sixth formers studying mathematics at GCE Advanced Level, some of whom will become future teachers of the subject, would have acquired 'positive' attitudes on these scales. The study showed however that there was a slight tendency for the A-level mathematician as compared with younger pupils to see:

> mathematics as more fixed and 'given once and for all'; as more difficult to learn and as having a lesser part to play in contemporary society. He also has less enthusiasm for school and is more inclined to the view that man is to some extent a slave of his environment. (Pidgeon, 1967, p. 276).

It is unlikely that attempts to monitor nationally a substantial part of the curriculum will be entirely successful in the first instance. In the United States the NAEP (National Assessment of Educational Progress) first formulated plans for monitoring pupils' performance in 1964. The first tests were administered in 1969 to samples of pupils and young people of age 9, 13, 17 and within the 26 to 35 age range. Currently some ten areas of the curriculum, including mathematics, have been monitored – some for the second time. In such cases there has been a rethinking underlying the testing programme and the quality of the material has been improved. The mathematics tests are criterion-referenced which demanded a positive move to clarify goals and objectives even though in some of the ten areas it apparently has been difficult to define other than trivial objectives. An interesting development has been the additional tape recording of test items so that individuals or small groups hear as well as see the questions – an attempt to remove any differences that could be attributed to reading skills.

In England, Wales and Northern Ireland the APU (Assessment of Performance Unit) was set up in 1975 with the aim of providing information about pupils' general levels of performance and how these change over the years. Bell (1977) states that:

> Later the work was given impetus by the realisation by the DES that, faced with allegations from industry of poor mathematical skills of school leavers, it had no objective and quotable evidence on which to base either graceful acceptance or a denial of these allegations. (p. 24).

The terms of reference of the APU are:

> To promote the development of methods of assessing and monitoring the achievement of children at school, and to seek to identify the incidence of under-achievement. (DES, 1980b, p. xi.).

The model of the curriculum for the purposes of assessment would be along six lines of pupils' development – language, mathematical, scientific, aesthetic, personal and social, and physical. It is understood that development in mathematics is not confined to the subject but extends to any subject in the school curriculum where many non-verbal forms of communication are used; it is also understood that, for example, language, scientific and aesthetic development may be assessed within the subject mathematics. It is therefore clear that the model stresses process and attitude rather than knowledge, skills and concepts – but not to the exclusion of the latter. The APU mathematics monitoring team, set up in September 1976, was able to build on the work of the TAMS (Tests of Attainment of Mathematics in Schools) project (Sumner, 1975; Kyles and Sumner, 1977). A useful introduction to the work of the APU in terms of the thinking of the mathematics team, prior to the first survey in 1978, can be found in Bell (1977).

Monitoring was carried out in 1978 on 13,000 eleven-year-old pupils in England and Wales and 14,000 fifteen-year-old pupils in England, Wales and Northern Ireland. Reports of these surveys can be found in DES (1980b, c) respectively. Surveys are to be repeated annually over a period of five years – in this way it is hoped to build up a study which is both cross-sectional and longitudinal. The report of the second survey of 14,500 eleven-year-old pupils in England, Wales and Northern Ireland can be found in DES (1981).

The surveys of eleven-year-old pupils comprised written (short answer) and practical mathematics tests and an attempt to assess attitudes towards mathematics. For the purposes of the testing, mathematics was divided by a two-way classification into five main content areas – geometry, measures, number, algebra and probability and statistics – and into six learning outcomes. Three of these outcomes related to products of learning – concepts, skills and applications; three related to the processes of doing mathematics – generalisation, proof and investigation. The practical tests were administered on a one-to-one interview basis to a sub-sample of 1,000 pupils. These tests, which sampled a restricted number of topics, were designed to assess only knowledge and skills in the first survey. There was greater stress on the exploration of pupils' reasoning and understanding in the second survey. The attitude surveys were administered to a different sub-sample of 1,500 pupils. The results of the first survey show that eleven-year-olds can do mathematics involving the fundamental concepts and skills and simple applications of them. However a sharp decline in performance was observed as pupils' understanding of concepts is probed more deeply and basic knowledge is applied in more complex situations or unfamiliar contexts. Pupils showed a general understanding of the basic ideas of symbols, graphs and diagrams. However it was found that some pupils find translating and manipulating symbols too abstract; how abstract depended on the familiarity of the symbol or its verbal translation. The following are taken from DES (1980b, pp. 130–1): *of* was more easily understood than X and *how many halves in two and a half* more easily understood than $2\frac{1}{2} \div \frac{1}{2}$; the reverse was true in the case of *what is the sum of* for + and *what is 5 times as big as 2* for 5×2. As far as attitudes are concerned there appeared to be some relationship between a liking for mathematics and its perceived difficulty. Pupils appear to have very clear ideas about the utility of the subject (usually represented by the 'four rules'), which did not necessarily match their perceptions of its difficulty. The second survey produced only few statistically significant differences from the first survey. These might be attributed to random fluctuations and so no conclusions about trends are offered. Attitudes were much the same as for the first survey. Boys were generally more self-confident about their ability in mathematics, whereas there appeared to be little difference between boys and girls as regards their perceived enjoyment of mathematics or of its utility.

In the survey of fifteen-year-old pupils, 10 per cent of the items were the same as or directly comparable to items used in the surveys of eleven-year-olds, with between 15 and 30 per cent more of fifteen-year-olds obtaining correct answers. No attempt to assess pupils' attitudes was made in this survey; an attempt to do so will be made in the second survey. In the written tests up to half of the pupils omitted items in the algebra category – specifically those involving quadratic equations, simultaneous equations, matrices and functions – and also in the geometry category – specifically those items on vectors and transformations. The omission of responses to the 'modern' topics may be an indication that many pupils are being taught on 'traditional' syllabuses. One result which seems to bear out the findings of cognitive research studies was in the sub-category *ratio* where on average only 31 per cent of pupils obtained correct answers to each of the items. Boys had higher scores in every one of the 15 sub-categories into which the five main categories were divided, and in eleven of these the difference was significant. This may be compared with the survey of eleven-year-old pupils where the differences were slight and boys were superior in ten of the 13 sub-categories. As regards the practical tests, the fifteen-year-old pupils needed more prompting to give correct answers than eleven-year-olds. In general it was found that fifteen-year-old pupils are not able to express mathematical ideas in words fluently. However in the practical tests virtually all pupils were familiar with a calculator and no pupil had any difficulty in operating the one provided.

Mathematics teachers may view the work of the APU with some apprehension because of the close association with the standards debate and discussions concerning the 'core curriculum'. On the positive side however we now have a base-line, which perhaps needs refinement, against which to compare the results of future APU surveys so that we may indeed determine whether or not standards really are falling and precisely in what respect. The encouraging feature of these surveys is that a deliberate start has been made to extend both the style of assessment and the range of instruments in use. In this respect teachers are urged to read the chapters on practical testing and on assessment of attitudes in each of the reports.

Mention should be made of recent surveys made by HMI. These were not exercises in monitoring attainment, not focused specifically on mathematics and not concerned primarily with assessment in schools. The surveys were conducted in order to identify and report on good practice in schools in terms of curriculum development and innovation, organisation and teaching methods across the whole range of school subjects.

The National Survey of Primary Education by HMI was carried out over the period 1975–7. The survey revealed that mathematics was given high priority by teachers to the extent that 88 per cent of schools used a scheme of work or a set of guidelines. For the purposes of the survey, the NFER administered one of their mathematics tests, E2, with results which did not seem to match the efforts of pupils (especially the more able), teachers and those involved with initial and in-service training. As a result HMI produced a report (DES, 1979a) which 'sets out to identify those points of the subject which, in HMI's view, should be taught to every child, and gives some guidance on approaches which teachers have shown to be effective.' (p. vii). However there is no attempt to identify minimum standards of pupil performance. In this report general aims are given for teaching mathematics and objectives, in terms of concepts and skills, for an undefined majority of pupils at eleven. Further objectives are suggested for some children. The main body of the report concentrates on the planning of the mathematics curriculum in the light of the suggested aims and objectives. As regards assessment it is recommended that schools should decide for themselves the purpose of assessment procedures – whether for grading purposes (which may be part of a local uniform scheme), to stimulate competition, for diagnostic purposes or as part of an evaluation procedure whereby teaching may be modified. The implication here is that a variety of types of test instruments and record keeping should be used; the onus is on the individual teacher to make the assessments.

The secondary survey report (DES, 1979b) was the outcome of visits to 384 schools in England by teams of HMIs. In the report one chapter is devoted to mathematics. Much of what is relevant to assessment is concerned with the influence of the public examination system on the school mathematics curriculum. HMI observed that 22 per cent of pupils in years four and five were committed to O-level courses (with no significant difference in the numbers of boys and girls), 13 per cent to O-level/CSE courses in the fourth year – although this figure was halved in the fifth year, and 52 per cent to CSE courses where there were slightly more girls than boys. The percentage of pupils on non-examination courses was variable as some schools had instituted limited-grade CSE Mode III courses for the least able pupils. The number of fourth and fifth-year pupils not studying any form of mathematics was very small. It is clear that parental and public demand for mathematics qualifications is very strong. HMI rightly point out that examination syllabuses are no substitute for systematic teaching schemes. In adopting an examination syllabus too rigidly teachers risk leaving out the more interesting topics which are not in the syllabus and giving less emphasis to those topics difficult to examine (p. 121). HMI deprecates work which is limited to the acquisition of low-level skills in order that pupils might answer narrow stereotyped questions (p. 121). It was found that, nationally, 20 per cent of the entries in mathematics for CSE examinations in 1976 were for Mode III schemes, which produce more varied patterns of assessment. Such variety poses problems in terms of moderation, but the pupils realised for themselves the value of course work and projects. HMI observed a wide divergence of practice in marking and assessment. Good marking would seem to encourage pupils and enhance their interest and morale (p. 145). In some schools the standard of presentation by some pupils was unsatisfactory and could usually be associated with poor standards of marking (p. 145). HMI recommend that teachers should adopt a policy of higher quality and more selective marking. Schools should give clear direction as to the purpose of marking in laying down a clear policy on marking and other methods of assessment (p. 146). The supplementary report on mathematics (DES, 1979b) contains further detail and also priority recommendations for improving the teaching of mathematics. As such the problems of assessment were not rated as high a priority as those areas specifically mentioned.

10.3 The Nature and Purposes of Assessment

It is necessary to make a distinction between *assessment* and *evaluation* as used in the educational context. Dictionary definitions are of little help since it would appear that these two words are almost synonymous. In the United Kingdom, these words have only relatively recently appeared in the educational context. Almost by common consent, but by no means with universal agreement, assessment has come to mean 'putting a value on pupil/student performance' – whether in terms of the products of learning, the strategies employed or the attitudes acquired. Evaluation has come to mean making a judgement on the effectiveness of the curriculum in terms of the suitability of the syllabus and how appropriate the teaching is. Hence we can say that we assess pupils but we evaluate curricula. It should be noted that in some literature originating from the United States the word evaluation is used in putting a value on student performance.

APU monitoring is almost exclusively focused on student performance, whereas the HMI surveys, mentioned in the previous section, are concerned primarily with evaluation – the traditional role of HMI being that of an evaluator. Assessment of student performance is necessarily part of the evaluation process. As was mentioned previously, some of the motivation to produce *Mathematics 5–11* (DES, 1979a) came from the results of assessing pupils using the NFER E2 test. In the Secondary Survey (DES, 1979b, 1980a), although no specific testing of pupils was carried out by HMI, much of what is said about mathematics in the reports relates to types of course as defined by examination objectives – the fact that pupils are so designated has an implication that some previous assessment of pupil performance had taken place.

In measuring pupil performance, which in itself is a rather vague term, is it attainment or achievement that is being assessed? There appears to be no set definition of either word which makes the difference obvious. It is up to the individual to produce his/her own definition as necessary. Sumner (1975 p. 12) suggests that attainment is a state of performance and achievement is a description of both the performance and how it comes about. As far as it is possible to do so the words attainment and achievement will be used with this very fine distinction in mind.

Assessment is more than the setting of written or practical tests or examinations, the marking of responses and the possible grading of pupils either into a simple rank order or by way of whether or not they have attained set criteria of performance. In mathematics, assessment may include making judgements about project work, investigations, aesthetic awareness, attitudes, etc. The style of assessment may range from the very formal use of a standardised, criterion-referenced or diagnostic test to the very informal use of question-and-answer technique and classroom observation. It may be the result of the very subjective feelings teachers have about a pupil's performance. There is a danger in that an excess of formalised testing occupies otherwise valuable teaching time. Teachers see the marking of pupils' work as being central to the assessment process. However, in the Secondary Survey, HMI found much to criticise in this aspect of assessment (DES, 1979b, pp. 145–6). There is a clear message here. Marking pupils' work is a common and effective means of assessing pupil achievement, and therefore it needs to be done carefully, selectively and meaningfully. Good marking can provide not only evidence of the product of learning but also of what mathematical strategies the pupil is using.

The very act of assessing pupil achievement assumes that the assessors have a very clear idea of what features of pupil performance they are seeking to assess – whether it be the outcomes of learning (products), the strategies employed in solving problems (processes) and/or the attitudes of pupils towards mathematics. Further, the assessor must be aware of what products, processes and attitudes should be assessed. 'It is all too easy to restrict assessment to these aspects of mathematics teaching which are most easily assessed.' (DES, 1979a, p. 10). It may therefore be necessary for teachers to devise new instruments to ensure that a more comprehensive picture of pupils' mathematical achievement can be obtained. Some of the work of the APU, reported by DES (1980b, c, 1981) could be used by teachers as a basis for developing their own new techniques.

Pupil behaviours are known as *objectives*, which should not in themselves be seen as absolutes. What objectives are seen as appropriate will depend on the reasons for teaching mathematics – whether determined by the individual teacher, departmental policy, school policy, by general agreement within an LEA, or nationally. These reasons are usually formulated as *aims* and 'give both direction and shape to a scheme of work or teaching programme.' (DES, 1979a, p. 5). Nationally or locally accepted aims are often supplemented by those agreed by a team within a school. Objectives are detailed descriptions of pupil performance by which aims are achieved. Having decided on aims and objectives, and the appropriate course of mathematics, the teacher or assessor must then determine appropriate styles and modes of assessment in order to ascertain whether the aims and objectives have been realised. The Schools Council (1979), in promoting a course entitled *Mathematics Applicable* which aims to make mathematics relevant and therefore interesting by immersing it in a modelling context, had also to promote a sustained modelling component in the terminal examination. Those mathematics teachers seeking guidance in the matter of choosing aims and objectives should consult DES (1979a, pp. 5-6 and pp. 77-9) and Mathematical Assocation (1976, pp. 2-6).

Assessment is not solely a matter of determining whether or not pupils have acquired certain products, processes and attitudes. In the matter of reporting the assessment one has to consider to whom the results are to be reported and for what purpose. There is not a great deal on either of these aspects that is contained specifically in mathematical education literature. The Mathematical Association (1974) suggests that assessments are made with the intention of reporting to some or all of teachers, pupils, parents, employers and further education (see p. 176). Rowntree (1977, pp. 16-33), although writing in a general context, lists six purposes of assessment: selection, maintaining standards, motivation of students, feedback to students, feedback to the teacher, preparation for life. There would seem to be three criteria for good assessment: (1) it should be *encouraging*; (2) it should *not be conditioned* by any external examination and (3) it should be in a form that is *useful* to pupils, parents, employers etc.

10.4 Some Issues in Assessment

The Case For and Against the Classification of Objectives

HMI, as a result of the Primary Survey, strongly recommended that schools/teachers firstly establish their aims for teaching mathematics, then translate these aims into descriptions of performance, objectives, before planning a detailed mathematics programme; subsequently to establish methods of assessment which could adequately measure which of the objectives have been realised by which pupils (DES, 1979a). In terms of public examinations at 16+ and 18+, many examining boards are attempting to specify behaviours to be assessed in written examinations; some CSE examining boards are even attempting to specify grade-related behaviours. The translation of aims into meaningful behaviours which adequately describe pupils' performance is not an easy task.

The impetus for structuring objectives by way of a 'scientific classification' came with the publication of volume one of a taxonomy of objectives by Bloom *et al.* (1956). This first volume was concerned solely with rational thinking skills i.e. in the cognitive domain. A second volume on the emotional feeling process, the affective domain, was published subsequently (Bloom *et al.*, 1964). The work of Bloom and his colleagues was not tied to any particular subject area, but a few of the illustrative examples are taken from mathematics. Much of the subsequent work in mathematics has concentrated on the classification of rational thinking, and the work of, for example, Husen (1967), Wood (1968) and Hollands (1972) is based very much on that of Bloom. The work of Avital and Shettleworth (1968) deserves particular mention – if only because of the apparent simplicity of their classification. They structure thinking skills into three levels: the lowest level – *recall* or *recognition* (equivalent to Bloom's category *knowledge*), the second level – *algorithmic thinking* (corresponding to Bloom's categories *comprehension* and *application*), and the highest level – *open search* (equivalent to Bloom's categories *analysis*, *synthesis* and *evaluation*). Such a classification would seem to have been cast within the context of problem solving procedures. As in all such classifications there is a hierarchy of levels such that each level subsumes all lower levels. Such classification hierarchies have been the basis for much of the work on classifying items used in objective tests (here the word objective is used in a different context – to describe the objective nature of the marking of pupils' responses). Such tests have become a feature of many GCE and CSE examinations in mathematics. The writing of items for objective tests is a skilled and time-consuming task. Wood and Skurnik (1969) set up an item-banking project of test items classified by content and by pupil behaviour in order to make available to mathematics teachers the possibility of constructing a school-based examination having national currency. There is disagreement as to the fine classification of higher-level thinking skills (those based on Bloom's categories comprehension, application, analysis, synthesis and evaluation), but most authors of classifications agree that there is clear distinction between knowledge (recall of basic facts, definitions etc.) and the higher level thinking skills.

One of the main drawbacks of establishing and using a classification of objectives as a basis for an assessment procedure is that there is an assumption that the assessor is completely aware of a pupil's previous experience. That, for example, an item classified as testing Avital and Shettleworth's open-search behaviour is testing that behaviour for all pupils being tested. It is possible that a particular item will test open search behaviours for one pupil and algorithmic thinking behaviours for another pupil whose pre-test experiences are likely to be different from the first. The fact that objectives are constructed in terms of observable behaviour cannot guarantee that any assessment made is a precise reflection of a pupil's internal mental activity. Such classifications may be convenient but are necessarily crude. The danger in formulating objectives prior to constructing the curriculum is in assuming that no other worthwhile objectives will be revealed as a result of the teaching/learning process, and it is thus possible that the teacher's view of learning is narrowed. Is it possible to construct objectives by which creativity might be assessed?

It is unfortunate that Bloom published cognitive and affective objectives separately for it might be assumed that these behaviours act independently. Aiken (1970), in reporting research carried out in the United States, suggests that an attitude is considered to be partly affective; that students who fail to achieve very much may develop negative attitudes and blame their teachers for their failures. Rees (1973) made this suggestion also. On the other hand Aiken found that teacher effectiveness had a positive effect on pupil attitude. Kempa and McGough (1977) investigated the attitudes of 300 first-year sixth formers in England towards mathematics and found that attitudes were strongly correlated with students mathematical bias as inferred from their choice of sixth-form subjects. The assessment of attitudes is an area where it is claimed there are no valid measures. In claiming that we are assessing cognitive behaviours can we be sure that we are not also assessing an affective behaviour also?

Process or Product?

Teachers' perceptions of mathematics vary from seeing the subject as a collection of established knowledge, skills and routines to be learnt, on the one hand, to a way of thinking about solving problems on the other. The former view implies a passive receiving role in learning; the latter view an active creative role. Of course one cannot solve problems without also having acquired knowledge, skills and understanding of concepts. Such views of mathematics have implications for the assessment of pupils' achievement. Thus teachers who see mathematics as a collection of facts to be learnt will wish to assess the products of learning; those who see mathematics as an activity will wish to assess the way pupils proceed in their use of mathematical strategies. Traditionally the narrow view of mathematics as solely a body of knowledge has been strong – but the current trend is towards a balanced view between active and passive learning. This trend began in primary schools and the balanced view is now well established there. In secondary schools we are only now witnessing the beginnings of a reform – despite the fact that a few teachers dedicated to the balanced view have been working in this area for at least 15 years (ATM, 1966). Thus assessment of pupil achievement in mathematics needs to reflect both process and product. Bell (1977) reports on the early thinking of APU in this respect. The work of the APU (DES, 1980b, c, 1981) almost guarantees that assessment of the process of learning mathematics will be given national standing.

The assessment of the products of learning mathematics, although a long practised art, is not without its difficulties. It is relatively simple to restrict such assessment to that which is most easily assessed (i.e. knowledge and skills). This in turn may severely restrict teaching to the acquisition of low-level skills. The assessment of pupils' learning strategies is a task that is more time consuming. Group and individual class activity can in themselves lead to increasing pressure on a teacher's time. It is all too easy when assessing pupils' work to ignore the methods used by successful pupils in order to devote more time to those who are in need of help (Jones, 1975). In his article Jones reveals that a great deal can be learnt about pupils' problem solving strategies by the methods they use to solve apparently simple problems. Researchers in pupils' cognitive development using interview techniques find that much can be revealed about pupils' apparent thinking and the strategies they employ. The practical tests using interview techniques employed by APU afforded a real opportunity to test a wider range of assessment procedures. An alternative method that is sometimes used is to require the pupil to provide a written commentary alongside his solution to a problem. The difficulty here is that many pupils, even at the age 16+, may not be capable of writing adequate descriptions of their thinking let alone appreciate the purpose for which they are doing it. Many of the findings of Hart (1981) are due entirely to the fact that interview techniques were used. The hard-pressed teacher may ask where can the time for interviewing individuals be found. The answer, inevitably, is it depends on what the teacher's priorities are and what evidence that is of value to the teacher can be so obtained. One of the teachers employed by APU as a tester wrote of practical mathematics 'It afforded an opportunity to hold a prolonged mathematical conversation with a child.' He added, 'My understanding of children's thought processes when solving problems has been considerably extended.' (DES, 1980b, p. 73).

Examinations or Course Work?

We shall distinguish between assessments derived from the formal, normally time-limited tests or examinations taken by pupils at specified times during a course and those assessments derived from the more informal pupil behaviour in completing practical work, investigations, essays and projects which are normally undertaken in a more relaxed and natural atmosphere. The setting of formal tests or examinations often produces a rather tense atmosphere which may for some pupils act as a distinct motivation and thus enable them to perform well. It is unlikely that this is true for many pupils and may well have the opposite effect for most. The Mathematical Association (1974, p. 177) points to the need to provide a more comprehensive pattern of assessment in order to assess those qualities such as creativity, imagination and sustained application which may not be well tested in formal written examinations. The sole use of written examinations to assess pupils' attainment introduces bias in the sampling of pupils' mathematical abilities. Consequently any attempt to draw conclusions about a pupil's general mathematical ability may not be valid. Examinations can test skills, concepts and applications very reliably. However teacher-made, end of term or year examinations may be of doubtful reliability and there is also a danger that their construction may be based on what has been taught rather than what has been learnt. Test construction is a task that requires much skill and expertise. The purpose for which a test is to be used is important. Grading pupils within a year group, checking mastery of particular skills or concepts or diagnosing individual pupils' weaknesses are purposes which require differently constructed tests. Reliable and valid norm-referenced (standardised), criterion-referenced and diagnostic tests are available commercially – both from the NFER and other publishers. The Mathematical Association (1979b) has produced a very useful booklet in an attempt to help teachers make informed choices in the matter of which, if any, tests to use. Every test should be issued with an official validity warning! Close attention must be paid to the content and purpose of the test and also to the age group and ability level of pupils for which the test was validated. HMI make this point (DES, 1979a). The correct age range is especially important as regards norm-referenced tests and in addition the age of the test (i.e. when norms were established). The gradual process of metrication and changes in mathematical notation may render some tests out of date. In order that the limitations of tests can be better appreciated it is important that teachers are aware how for example the NFER tests are constructed.

The assessment of investigations, practical work and projects is recognised as being difficult. Peck (1972) produces some course work submitted by a pupil for a Mode III CSE mathematics examination and challenges the reader to assess it. Fielker (1966) briefly discusses the difficulty of assessing investigations even when the assessor is experienced in this area. Neither writer really suggests a *modus operandus* for assessing investigations and leaves the reader with the feeling that assessment in this area must inevitably be impressionistic. The Mathematical Association (1980), pp. 55-8) does suggest a possible workable scheme, based on criteria, of assessment of project work. Although the scheme was devised for a particular situation it could well be adapted more generally to investigations as well as project work. Marsh (1975) discusses the reliability of estimates of attainment as revealed by practical situations in mathematics, when pupils are tested in a one-to-one situation. This study formed part of the TAMS (Tests of Attainments in Mathematics in Schools) project – the work of which formed the basis for that of the APU mathematics monitoring team.

There are two problems associated with the assessment of course work. The first is how does the teacher reconcile the possibly conflicting roles of both tutor and assessor. As tutor the teacher is expected to give some help and guidance to the pupil; as assessor he must judge the worth of the pupil's effort whilst recognising his the teacher's help. Many would argue that this is the ideal situation for an assessor to be in. The second is how to compare one piece of course work with another which may differ markedly in scope, presentation and style. A scheme of assessment based on criteria agreed by a group of teachers might overcome these problems. Even the assessor who bases his judgement on impression is using criteria which have yet to be formalised in writing.

Continuous or Terminal Assessment?

This is essentially a different issue from that discussed in the previous section. Assessment of course work is not necessarily a form of continuous assessment; examinations are not the only form of terminal assessment. Course work can be set and assessed at the very end of a learning sequence or course of study. Examinations or tests can be used effectively during a learning sequence to monitor pupils' performance over a period of time. It is possible that it is only appropriate to assess students at the end of a course of study. Such terminal, or summative, assessment admits no feedback which could alert the teacher to a necessary change in tactics during the period or provide feedback of help to the students. Any feedback obtained can only benefit the following cohort of pupils. Of course all teachers do make informal assessments through the use of observation, question-and-answer technique and marking pupils' work. What is meant by continuous, or formative, assessment is the deliberate attempt to measure pupils' achievement using mastery tests, diagnostic tests and/or assessment of practical work, projects and investigations. Continuous assessment is regarded as crucial by some teachers since upon such information they are able to make the necessary adjustments relevant to the process of evaluating their own teaching methods. Pupils need continuous feedback as to their progress in order to identify their strengths and weaknesses and to confirm or otherwise their present understanding or approach.

Rowntree (1977) identifies the real problem when he suggests that the issue is not one of assessment but of grading – continuous grading or terminal grading. The use of terminal grading, based on terminal assessment, is subject to errors due to the infrequent sampling of pupils' abilities. Continuous grading, based on both continuous and terminal assessments, is likely to reflect on a more comprehensive picture of a pupil's achievement.

Externally Certificated Assessment – School Based, External or Both?

The discussion which follows is certainly directed towards teachers in secondary schools, but may be equally pertinent for teachers in primary and middle schools. Many LEAs have devised tests of mathematical attainment to be administered at the stage of transfer in order to provide information intended to help teachers in the next stage. In the Secondary Survey (DES, 1979b) HMI found that, in the schools surveyed, 87 per cent of fourth-year pupils and 83 per cent of fifth-year pupils were following courses leading to GCE O-level or CSE mathematics examinations (p. 116). In 1976, of the total national entries for CSE mathematics examinations, only 20 per cent were for Mode III examinations (p. 119). Thus the majority of secondary school pupils in their fourth and fifth years are following courses leading to an externally set, externally assessed examination. Cohen and Deale (1977) carried out a survey of the extent of assessment and moderation by teachers in GCE and CSE examinations. As regards CSE Mode I examinations in mathematics only two examining boards did not have a teacher assessed component. For ten boards the amount of teacher assessment varied from 10 per cent to 50 per cent of the total assessment. For the remaining two boards initial marking of *all* work was done by teachers working in a consortium. The GCE examining boards have no teacher assessment in Mode I type mathematics examinations, and Mode III type mathematics examinations are a rarity.

In the proposed Common System of Examining at 16+ none of the operational pilot schemes in mathematics included project work; only one study included teacher assessment of course work (Brown, 1975). The proposals promise that Mode III examinations will be available as under the present dual system.

The common arguments for and against both school based and external assessment are well known. External assessment is impartial, based on large samples of pupils, maintains 'accepted' standards and is performed by 'experts'. School based assessment provides the opportunity to sample abilities not appropriately assessed by other means, can produce a more comprehensive picture of a pupil's abilities by sampling them over a period of time and is performed by an assessor who 'really knows the pupil'. There are other equally pertinent arguments. Success in external examinations at 16+ and 18+, especially in mathematics, gives a pupil a 'licence' to enter certain professions or to continue education at the next level. Employers are still suspicious of CSE Mode III examinations, possibly because they fail to appreciate the need for special syllabuses and alternative methods of assessment to meet the needs of special groups of pupils, not necessarily of low ability, despite the fact that the grades are externally moderated. There lies another problem – the difficulty of moderating widely disparate forms of assessment and in deciding how much support the pupil has received in the presentation of project and investigation reports. External assessment has a restricting effect both in terms of what is taught and how it is assessed. For example, the fact that questions involving the use of the four rules with numbers in bases other than ten have in the past appeared in GCE O-level and CSE examination papers, has tended to mislead teachers as to the purpose and real value of so called 'multi-base arithmetic'. Most teachers involved in preparing pupils to take external examinations feel it is their duty to get their pupils to practice the art of taking examinations. Kay (1977) warns of three possible consequent effects – the adoption of short-cut methods in an effort to 'beat the examiner', neglecting those parts of curriculum difficult to test and therefore not tested, and concentrating on those pupils likely to achieve the minimum objectives. We might add further that methods of assessment not employed in the external examination may never be used within the teaching programme. More importantly pupils also tend not to see the point of learning material or submitting to assessment procedures not tested or employed in the external examination.

The debate concerning the proposals for a Common System of Examining at 16+ is likely to be focused most strongly on the establishment of national criteria. The present system which is essentially norm-referenced is likely to be replaced by a system which will be partially, if not totally, criterion-referenced. The general acceptance of certain performance criteria may imply a common curriculum or render the choice of syllabus irrelevant. It will depend on the detail with which the criteria are specified. The choice of criteria will almost certainly determine the methods of assessment that are appropriate. We could find ourselves no further from where we are at present with the main emphasis on written examinations. On the other hand mathematics teachers could be positively encouraged to use projects, investigations and practical work. In the Secondary Survey, HMI found that for GCE courses only 9 per cent contained an experimental/practical element, 5 per cent used topic/project work and 11 per cent allowed opportunity for individual enquiry. The corresponding figure for CSE courses were 20, 24 and 12 per cent (DES, 1980a, p. 29).

Many teachers are apprehensive about the idea of teaching against a background of performance-criteria. Taking a criterion-referenced test is not a new experience for most people since there are well known examples outside the school context, for example grade examinations in music and the driving test. The latter is also diagnostic test since those who fail obtain some feedback as to the particular skills they lack facility in. Some critics of both the present and proposed systems would go as far as suggesting that public examinations at 16+ should be abolished and replaced by criterion-referenced tests at various levels which could be taken by any pupil at any age who had been suitably prepared and was ready to take them; a system very similar to the grade examinations in music which contain both theoretical and practical elements.

Reporting Assessments – Grades or Profiles?

This would seem to be the most crucial issue in assessment generally. Are single grade letters, numbers or percentage marks adequate descriptions of a pupil's performance? This question is particularly relevant when the results from differing forms of assessment, used to assess the widely differing abilities of an individual, are aggregated to provide a single descriptor. Single grade numbers or percentage marks are susceptible to statistical analysis – a process which may be completely inappropriate. Research shows that there is unreliability in the final gradings awarded in GCE and CSE examinations. A candidate's 'true' grade may be within one grade either side of that awarded. Hughes (1971) discusses the Schools Council's proposals for a system of grading in A-level examinations which recognised the uncertainty in examination results by suggesting that grades be given within a range of uncertainty – thus warning users of grades not to attach a greater degree of accuracy to these results than lies in the statement. Alas, even this step in the right direction was by-passed. Bishop (1975) questions the validity of research which makes use of the statistical analysis of CSE grades, which are norm-referenced. (Norm-referencing only places the candidates in an order of merit and reveals little about the abilities of the candidates.) Bishop goes on to discuss some of the problems of aggregating those grades or marks awarded for different components of an examination into a single overall grade. Measures of different abilities should never be aggregated unless there is a high degree of correlation between them. Even when the various components supposedly measure the same abilities the method of weighting each component in the overall aggregation of marks will determine a particular rank order of candidates (see Forrest, 1974). Inevitably, in the process of aggregation some detail of performance which may reveal much about a candidate is lost. Many employers wrongly believe that the examination system provides sufficient information to enable them to choose the candidate appropriate for the job.

At any stage of transfer between one stage of education and the next there is a problem of what information should be passed on and in what form. Information from a pupil's previous school may be limited and in mathematics of little value to the new school. Knights and Waite (1974) discuss the problems of drawing up mathematical profiles for first-year secondary pupils. Ideally, profiles should not be general statements of achievement across a range of abilities but should reveal what a pupil can do. Norm-referenced tests or examinations would not seem suited to this purpose. Criterion-related performance in written examinations should be capable of being reported in profile form. To such a profile could be added aspects of pupil performance in practical work, investigations and assignments. Many examining boards claim they do not have the data-processing power to provide profiles as opposed to single-grade descriptors. Perhaps the way forward is for schools to construct pupil profiles, and for those pupils at the 16+ and 18+ stages their profiles could include components externally assessed. What about pupil attitudes – are these not as relevant as assessments of 'product' and 'process'? The research by Preston on the measurement of affective behaviour in CSE mathematics, reported in Fairbrother (1980, pp. 120–5), includes an interesting example where two pupils achieved identical gradings in the CSE examination but for whom different mathematical attitude profiles could be constructed. It is dangerous to generalise from one particular example, but many employers would claim an interest in a prospective employee's motivation, interest and commitment – as would many teachers.

There is little reported research into the construction and use of profiles in school mathematics. The Mathematics in Education and Industry Committee of the Mathematical Association has produced a system to test the ability of school leavers to demonstrate basic numerical skills. Pupil attainment is reported in profile form – (16+)

School Leavers Attainment Profile of Numerical Skills (SLAPONS). Each pupil thus tested obtains a profile of attainment which can be matched against an employers profile of 'minimum attainment for the job'. The impetus for this work came from schools' general dissatisfaction with employers' unseen tests – often out-of-date and home-made. The main drawback seems to be in the interpretation of a profile where the pupil's performance in some areas falls short of and in other areas exceeds that expected by employers. The main virtue of this interesting exercise seems to be in getting teachers and employers together to discuss a very important problem; for teachers to consider which numerical skills it is important for school leavers to be proficient at; for employers to think carefully about the numerical skills required for the job. The Mathematical Association (1979a) has given some thought to the problem of reporting assessments and makes suggestions, which are relevant to both primary and secondary teachers, about attitude assessments (pp. 15-17), for improved record keeping generally (pp. 23-30) and for more detailed attainment profiles (pp. 31-40).

10.5 Assessment and Its Relation to the Evaluation of School Mathematics

The Mathematical Association (1979a) suggests that:

> The process of evaluation is a complex one including: assessment of children's performance; evaluation of the process of teaching and its effect on children's ability to learn; evaluation of the curriculum being followed, its suitability to the needs of the children and the expertise and style of the teacher teaching it. (p. 41)

One of the criteria by which the success of a school is judged by those outside is the performance of its pupils in public examinations. In this respect mathematics is an important feature. Thus a school's mathematics curriculum may appear to be highly successful with nothing said about the relevance of the curriculum, the quality of the teaching and the interest shown by the pupils. What the pupils fail to achieve may be far more important than what is achieved. The use of terminal assessment can only contribute to what is termed summative evaluation. Considered alone such evaluation at the end of a course alerts the teacher too late if and when the curriculum or teaching methods are not entirely appropriate for a particular group of pupils.

At best it is only a crude indicator that something needs to be done to improve the learning situation for the following cohort of pupils. What seems to be necessary is some form of ongoing or formative evaluation by which the teacher can continuously apply 'fine tuning' to the content, teaching style and processes in order to maximise their suitability to the learning needs of pupils. HMI, in their report of the Primary Survey (DES, 1979a), stated clearly that in their opinion 'If teaching is to be successful, it is essential that the teacher should assess what is happening.' (p. 9), and again 'It is necessary to evaluate what both individuals and groups are learning; the results may or may not reflect what the teacher believes she has taught.' (p. 10). Such on-going assessment is usually performed subjectively through observation, oral questioning and discussion with individual pupils, and perhaps more objectively through marking, the use of specialised tests and assessment of course work.

Assessment marks or grades which are norm-referenced or place pupils in a rank order of merit are of little value in the evaluation process. In such cases teachers know which pupils are better than others but not necessarily what they are good at, unless they are prepared to analyse a pupil's responses more deeply. Rowntree (1977, pp. 181–2) quotes very aptly from Lewis Carroll:

> '. . . how can you possibly award prizes when everybody missed the target?' said Alice. 'Well,' said the Queen, 'some missed by more than others and we have a fine normal distribution of misses, which means we can forget about the target.'

It would thus seem that as far as objective evaluative measures are concerned we need criterion-related assessment of both mathematical content and processes. The results of an assessment of pupils performance should either encourage the teacher to use very similar methods another time or direct him to reassess his approach and teaching methods, the level of difficulty of the work and the suitability of topics and their presentation. Very often a team or departmental approach incorporating a variety of views and emphases can be helpful.

In the discussion so far there has been no reference to evaluation by pupils. Inevitably the results of teacher assessment, if revealed, enable pupils to self-assess their own abilities and, in addition, form perceptions of teachers' expectations in terms of criteria of performance. It should also be remembered that pupils evaluate their teachers in terms of their teaching skill and knowledge of, attitudes to and interest in mathematics. As a consequence, pupils develop positive or negative attitudes to mathematics. Aiken (1970) found that, in the United States, children's attitudes towards arithmetic were less dependent on organisation than on teachers' attitudes and the teaching methods they employ. As a consequence of self-assessment of their own mathematical achievement and evaluation of the teaching they have received, pupils arrive at an evaluation of mathematics itself. Their answers to questions such as 'Is mathematics difficult to learn, is it enjoyable to do, is it interesting, is it satisfying, is it useful?' will reveal much about whether we as teachers have achieved our aims in teaching mathematics.

Bibliography

Aiken, L.R. (1970) 'Attitudes towards Mathematics', *Review of Educational Research*, *40*, 551–96.
ATM (Association of Teachers of Mathematics) (1966) *The Development of Mathematical Activity in Children; the Place of the Problem in this Development*, ATM, Nelson, Lancs.
Avital, S.M. and Shettleworth, S.J. (1968) *Objectives for Mathematics Learning*, The Ontario Institute for Studies in Education, Ontario.
Bell, A. (1977) 'The APU and the 1978 Mathematics Survey', *Mathematics Teaching*, No. 80, pp. 24–7.
Bishop, A. (1975) 'Examinations and What They Examine', *Mathematics Teaching*, No. 70, pp. 44–5.
Bloom, B.S. (ed.) (1956) *Taxonomy of Educational Objectives, Handbook 1, The Cognitive Domain*, Longmans, London and New York.
——— (ed.) (1964) *Taxonomy of Educational Objectives, Handbook 2, The Affective Domain*, Longmans, London and New York.
Brown, M. (1975) 'The 16+ Exam', *Mathematics in School*, Vol. 4, No. 2, pp. 5–7.
Cohen, L. and Deale, R.N. (1977) *Assessment by Teachers in examinations at 16+*, Schools Council Examinations Bulletin 37, Evans/Methuen, London and New York.
DES (Department of Education and Science) (1979a) *Mathematics 5-11: A Handbook of Suggestions*, HMSO, London.
——— (1979b) *Aspects of Secondary Education in England: A Survey by HM Inspectors of Schools*, HMSO, London.
——— (1980a) *Aspects of Secondary Education in England: Supplementary Information on Mathematics*, HMSO, London.
——— (1980b) *Mathematical Development: Primary Survey Report No. 1*, HMSO, London.
——— (1980c) *Mathematical Development: Secondary Survey Report No. 1*, HMSO, London.
——— (1981) *Mathematical Development: Primary Survey Report No. 2*, HMSO, London.
Fairbrother, R.W. (ed.) (1980) *Assessment and the Curriculum*, Chelsea College, University of London.
Fielker, D. (1966) 'Assessing Investigations' *Examinations and Assessment*,

pp. 68–70, Mathematics Teaching Pamphlet number 14, Association of Teachers of Mathematics, Nelson, Lancs.

Forrest, G.M. (1974) 'The Presentation of Results' in MacIntosh, H.G. (ed.) *Techniques and Problems of Assessment*, Edward Arnold, London.

Hart, K.M. (ed.) (1981) *Children's Understanding of Mathematics: 11-16*, John Murray, London.

Hollands, R. (1972) 'Educational Technology. Aims and Objectives in Teaching Mathematics' *Mathematics in School*, Vol. 1, No. 2, pp. 23–4; No. 3, pp. 32–3; No. 5, pp. 20–21; No. 6, pp. 22–3.

Hughes, M. (1979) 'Testing Numeracy', *Mathematics Teaching*, No. 89, pp. 28–9.

Hughes, S. (1971) 'A new "A" level grading system', *Secondary Education*, Vol. 1, No. 3, pp. 23–6, National Union of Teachers, London.

Husen, T. (1967) *International Study of Achievement in Mathematics – Vols. I and II*, John Wiley, London and New York.

IMA (Institute of Mathematics and its Applications) (1976) 'A National Standard of Basic Skills in School Mathematics', *Bulletin*, *12*, 320–1, IMA.

——— (1977) 'A Proposal for A National Standard of Basic Skills in School Mathematics', *Bulletin*, *13*, 66–7, IMA.

——— (1978) 'A Pilot Test of Basic Numeracy of Fourth and Fifth-year Secondary School Pupils undertaken by the Institute', *Bulletin*, *14*, 83–6, IMA.

Jones, D.A. (1975) 'Don't Just Put the Answer – Have a Look at the Method!', *Mathematics in School*, Vol. 4, No. 3, pp. 29–31.

Kay, B.W. (1977) 'Monitoring: Purpose, Desirability and Scope', in Sumner, R. (ed.) *Monitoring National Standards of Attainment in Schools*, pp. 14–26, NFER, Slough.

Kempa, R.F. and McGough, J.M. (1977) 'A Study of Attitudes towards Mathematics in relation to Selected Student Characteristics', *British Journal of Educational Psychology*, *47*, 296–304.

Knights, G. and Waite, P. (1974) 'Assessing the Mathematical Abilities of Children', *Mathematics Teaching*, No. 68, pp. 16–18.

Kyles, I. and Sumner, R. (1977) *Tests of Attainment in Mathematics in Schools – Continuation of Monitoring Feasibility Study*, NFER, Slough.

Marsh, J. (1975) 'A Feasibility Study in Assessing Practical Maths', *Mathematics in School*, Vol. 4, No. 6, pp. 21–2.

Mathematical Association (1974) *Mathematics Eleven to Sixteen*, Bell, London.

——— (1976) *Why, What and How?*, The Mathematical Assocation, Leicester.

——— (1979a) *Evaluation: Of what, by whom, for what purpose?*, The Mathematical Association, Leicester.

——— (1979b) *Tests*, The Mathematical Association, Leicester.

——— (1980) *Pupils' Projects. Their use in Secondary School Mathematics*, The Mathematical Association, Leicester.

McIntosh, A. (1979) 'When Will They Ever Learn?' *Mathematics Teaching*, No. 85, Primary Supplement, pp. i–iv.

Peck, D. (1972) 'Assessment Workshop. How do you Assess Coursework?' *Mathematics in School*, *1*, pp. 12–16.

Pidgeon, D.A. (1967) *Achievement in Mathematics*, NFER, Slough.

Rees, R.M. (1973) *Mathematics in Further Education: Difficulties experienced by craft and technician students*, Brunel Further Education Monograph 5, Hutchinson, London.

——— (1976) 'Mathematics in Teacher Training Institutions: Some Difficulties Experienced by Teachers in Training', *Bulletin*, *12*, 328–32, Institute of Mathematics and its Applications.

Rowntree, D. (1977) *Assessing Students. How Shall we Know Them?*, Harper &

Row, London and New York.
Schools Council (1979) *Teaching Mathematics Applicable. Introductory Guide*, Heinemann, London.
Sturgess, D. (1978) 'What *Should* They Know at Sixteen?', *Mathematics Teaching*, No. 85, pp. 22–4.
Sumner, R. (1975) *Tests of Attainment in Mathematics in Schools – Monitoring Feasibility Study*, NFER, Slough.
Wood, R. (1968) 'Objectives in The Teaching of Mathematics', *Educational Research*, *10*, pp. 83–98.
Wood, R. and Skurnik, L.S. (1969) *Item Banking*, NFER, Slough.

Further Reading

Fraser, W.G. and Gillam, J.N. (1972) *The Principles of Objective Testing in Mathematics*, Heinemann Educational, London.
MacIntosh, H.G. (1974) *Techniques and Problems of Assessment*, Edward Arnold, London.
Montgomery, R. (1978) *A New Examination of Examinations*, Routledge & Kegan Paul, London.
NCTM (National Council of Teachers of Mathematics) (1961) 26th Yearbook, *Evaluation in Mathematics*, NCTM, Washington.
Satterley, D. (1981) *Assessment in Schools*, Blackwell, Oxford.
Wynne-Willson, W.S. (1978) 'Examinations and Assessment', in Wain, G.T. (ed.) *Mathematical Education*, Van Nostrand Reinhold, Wokingham and New York.

CONTRIBUTORS

Rolph Schwarzenberger is Professor of Mathematics and Chairman of the Science Education Department at the University of Warwick.

Peter Reynolds is Mathematics Adviser for Suffolk.

Michael Cornelius is a lecturer in the School of Education, University of Durham.

Peter Richards is a lecturer in the School of Education, University of Bath.

Derek Woodrow is Head of the Department of Home Economics, Mathematics and Science at the City of Manchester College of Higher Education.

Renée Berrill is a lecturer in the School of Education, University of Newcastle upon Tyne.

Colin Noble-Nesbitt is Head of Mathematics at St Leonard's Comprehensive School, Durham.

John Dunford is Deputy Headmaster at Bede School, Sunderland.

Keith Selkirk is a lecturer in the School of Education, University of Nottingham.

David Carter is a lecturer in the School of Education, University of Leeds.

GLOSSARY

This brief guide to some educational terms may be helpful to readers who are not familiar with the educational system in England.

Assessment of Performance Unit (APU): Set up in 1975 by the Department of Education (q.v.) 'to promote the development of methods of assessing and monitoring the achievement of children and to seek to identify the incidence of under-achievement'.

Association of Teachers of Mathematics (ATM): A national association, founded in 1952, which publishes a journal *Mathematics Teaching*, occasional books and a variety of materials developed by local groups.

Certificate of Secondary Education (CSE): An examination normally taken by pupils at the age of 16, aimed broadly at pupils between the 20th and 60th percentiles in the ability range. It is substantially run by teachers and there are three modes of examination. In *Mode 1* schools submit candidates for the regional examining boards' examinations; in *Mode 2* schools or groups of schools put up their own examination schemes for approval by the examining boards; and in *Mode 3* the examinations are set and marked internally by individual schools and moderated by the boards.

Cockcroft Report: The report of the Committee of Inquiry into the Teaching of Mathematics in Schools under the chairmanship of Dr W.H. Cockcroft set up by the Government in March 1978. The report *Mathematics Counts* was published in January, 1982.

Comprehensive School: Secondary school (q.v.) which takes all the children from a particular neighbourhood.

Concepts in Secondary Mathematics and Science: A research programme based at Chelsea College, London from 1974–79.

Department of Education and Science (DES): The Government department responsible for education in England and Wales.

'Fletcher' Mathematics: A series of books, much used in English primary schools (q.v.), originally written largely by Harold Fletcher and published by Addison-Wesley.

General Certificate of Education (GCE): Certificate awarded on the results of external examinations set by examining bodies in England and Wales. There are two levels: (1) *Ordinary (O) Level* normally taken at 16 and (2) *Advanced (A) Level* normally taken two years after O-Level.

Grammar School: Secondary school (q.v.) for the most able pupils selected at the age of eleven. Now largely replaced by comprehensive schools (q.v.).

Her Majesty's Inspectors (HMI): Group of officials attached to the Department of Education and Science (q.v.) who inspect schools and make reports.

Local Education Authority (LEA): There are 104 of these in England and Wales and they are responsible for the education in schools and colleges within their boundaries.

Institute of Mathematics and its Applications (IMA): A professional institution for mathematicians which organises conferences and publishes a Journal and a Bulletin.

Mathematical Association: A national association founded in 1871. It publishes reports on the teaching of mathematics and two journals the *Mathematical Gazette* and *Mathematics in School*.

Mathematics for the Majority Project: A project set up in 1967 with the aim of producing teacher guides for use with 13- to 16-year-old pupils of average and below-average ability.

Middle School: A school for children aged eight or nine to twelve or thirteen.

National Foundation for Educational Research (NFER): A body funded by the Government and local education authorities (q.v.) to carry out research into all areas of education.